I0484124

CULVERT DESIGN FOR AQUATIC ORGANISM PASSAGE

1. Report No. FHWA-HIF-11-008 HEC-26	2. Government Accession No.	3. Recipient's Catalog No.	
4. Title and Subtitle Culvert Design for Aquatic Organism Passage Hydraulic Engineering Circular Number 26		5. Report Date October 2010	
		6. Performing Organization Code	
7. Author(s) Roger T. Kilgore, Bart S. Bergendahl, and Rollin H. Hotchkiss		8. Performing Organization Report No.	
9. Performance Organization Name and Address Kilgore Consulting and Management Denver, Colorado		10. Work Unit No. (TRAIS)	
		11. Contract or Grant No.	
12. Sponsoring Agency and Address Central Federal Lands Highway Division 12300 West Dakota Ave. Lakewood, Colorado 80228		13. Type of Report and Period Covered Final Report 11/1/07 – 11/30/09	
		14. Sponsoring Agency Code	

15. Supplementary Notes

Contracting Officer's Technical Representative: Bart Bergendahl
Technical Support: Christopher Goodell, Alan Johnson, Philip Thompson, Steve Rainey, Christopher Frei

16. Abstract

This document presents a stream simulation design procedure, methods and best practices for designing culverts to facilitate aquatic organism passage (AOP). The primary goal of this document is to incorporate many of the current geomorphic-based design approaches for AOP while providing a procedure based on quantitative best practices. It presents a bed stability-based approach that accounts for the physical processes related to the natural hydraulic, stream stability, and sediment transport characteristics of a particular stream crossing. Specific information on fish, or other aquatic organisms, is not required, but should be incorporated when required.

The document provides a context for stream crossing design and describes the applicability of the design procedure. It also provides important background information a designer should be familiar with including how culverts create barriers, techniques for culvert assessments and inventories, fish biology, fish passage hydrology, stream geomorphology, construction, and post-construction. Detailed technical information supporting the practices used within the design procedure and several design examples are included in the appendices.

The core of the document is a 13-step design procedure. Step 1 involves determination of the hydrologic requirements for the site for both flood flows and passage flows. Step 2 defines the project reach and establishes the representative channel characteristics appropriate for the design. Steps 3 and 4 are to identify whether the stream is stable (Step 3). If not, channel instabilities are analyzed and potentially mitigated (Step 4). In Step 5, an initial culvert size, alignment, and material are selected based on the flood peak flow. Subsequently, the stability of the bed material is analyzed under the high passage flow (Steps 6 and 7) and flood peak flow (Steps 8 and 9). Steps 10, 11, and 12 focus on the velocity and depth in the culvert. However, these parameters are not compared with species-specific values, but rather are compared with the values upstream and downstream of the culvert insuring that if an organism can pass the upstream and downstream channel, it will also be able to pass through the culvert. Step 13 allows the designer to review the completed design.

17. Key Words: Culvert design, Aquatic organism passage (AOP), Fish passage		18. Distribution Statement No restrictions. This document is available to the public through the National Technical Information Service, Springfield, VA 22161	
19. Security Classif. (of this report) Unclassified	20. Security Classif. (of this page) Unclassified	21. No. of Pages 234	22. Price

Form DOT F 1700.7   (8-72)   **Reproduction of completed page authorized**

ACKNOWLEDGMENT

Kilgore Consulting and Management (Roger Kilgore, P.E., Principal) prepared this first edition of HEC 26. In addition to Mr. Kilgore, the project team included Rollin Hotchkiss, Ph.D., P.E. (Brigham Young University), Christopher Goodell, P.E. (WEST Consultants), Alan Johnson (Aquatic Resource Consultants), Philip Thompson, P.E. (independent consultant), Steve Rainey, P.E. (Fish Passage Engineer, GEI Consultants), and Jeff Bradley, Ph.D., P.E. (WEST Consultants).

The work performed by Rollin Hotchkiss and Christopher Frei in FHWA's "Design for Fish Passage and Road-Stream Crossings: Synthesis Report" (2007) was used to identify a broad range of background material and references.

The work was completed under the direction of FHWA's Project Manager, Bart Bergendahl, P.E., who also contributed significantly to the content of this document.

Several reviewers from a Technical Advisory Committee and the FHWA contributed comments, corrections, and suggestions. These contributors were:

Technical Advisory Committee

Mike Furniss	United States Forest Service
Robert Gubernick	United States Forest Service
Charles Hebson	Maine Department of Transportation
Andrzej Kosicki	Maryland State Highway Administration
Bryan Nordlund	National Marine Fisheries Service
Mark Weinhold	United States Forest Service
Marcin Whitman	California Department of Fish and Game

FHWA

Larry Arneson	Hydraulics Engineer
Eric Brown	Hydraulics Engineer
Mark Browning	Hydraulics Engineer
David Carlson	Environmental Specialist
Stephen Earsom	Environmental Specialist
Paul Garrett	Ecologist
Mary Gray	Environmental Specialist
Scott Hogan	Hydraulic Engineer
Kevin Moody	Environmental Specialist
Melissa Schnier	Environmental Biologist

TABLE OF CONTENTS

                                                                                    Page
ACKNOWLEDGMENT...I
TABLE OF CONTENTS..II
LIST OF TABLES VIII
LIST OF FIGURES X
GLOSSARY XII
GLOSSARY OF ACRONYMS XVIII
LIST OF SYMBOLS..XX

CHAPTER 1 - INTRODUCTION 1-1

1.1   PURPOSE...1-1
1.2   CONTEXT 1-2
        1.2.1   Historical Crossing Design.....................................1-2
        1.2.2   Road Stream Interaction...1-2
1.3   DESIGN PROCEDURE APPLICABILITY 1-3
1.4   MANUAL ORGANIZATION 1-3

CHAPTER 2 - CULVERTS AS PASSAGE BARRIERS 2-1

2.1   STREAM FRAGMENTATION..2-1
2.2   BARRIER MECHANISMS 2-2
        2.2.1   Drop at Culvert Outlet..2-2
        2.2.2   Outlet Pool Depth 2-2
        2.2.3   Excessive Barrel Velocity 2-3
                 2.2.3.1   Boundary Layer Velocity 2-3
                 2.2.3.2   Average Velocity 2-4
                 2.2.3.3   Maximum Point Velocity 2-4
                 2.2.3.4   Inlet Transition Velocity 2-4
        2.2.4   Insufficient Depth..2-4
        2.2.5   Excessive Turbulence 2-4
        2.2.6   Culvert Length 2-5
        2.2.7   Debris and Sediment Accumulation..........................2-5
        2.2.8   Culvert Damage..2-5

CHAPTER 3 - AOP CULVERT ASSESSMENT AND INVENTORY 3-1

3.1   AOP CULVERT ASSESSMENT..3-1
        3.1.1   Assessment Criteria 3-1
        3.1.2   Degree of Barrier..3-1
        3.1.3   Data Collection 3-2
3.2   CULVERT INVENTORY 3-3
        3.2.1   Road-based Inventory 3-3
        3.2.2   Stream-based Inventory 3-4

TABLE OF CONTENTS (CONTINUED)

Page

CHAPTER 4 - FISH BIOLOGY **4-1**

4.1   CAPABILITIES AND ABILITIES 4-1
       4.1.1   Swimming and Jumping 4-1
       4.1.2   Species and Life Stages 4-2
       4.1.3   Depth 4-3
       4.1.4   Exhaustion 4-4
4.2   MIGRATION AND MOVEMENT 4-5
       4.2.1   Anadromous Fish 4-5
       4.2.2   Juvenile and Resident Fish 4-5
       4.2.3   Fish Presence 4-6

CHAPTER 5 - PASSAGE HYDROLOGY **5-1**

5.1   SEASONALITY AND DELAY 5-1
5.2   DESIGN HYDROLOGY 5-2
       5.2.1   Flood Peak, Q_p 5-2
       5.2.2   High and Low Passage Flows 5-3
5.3   FLOW DURATION CURVES 5-5

CHAPTER 6 - STREAM GEOMORPHOLOGY **6-1**

6.1   CHANNEL CHARACTERISTICS 6-1
       6.1.1   Channel Width 6-1
               6.1.1.1 Active Channel Width 6-1
               6.1.1.2 Bankfull Width 6-1
       6.1.2   Gradient 6-2
       6.1.3   Bed Material and Embedded Culverts 6-2
       6.1.4   Key Roughness Elements 6-2
6.2   CHANNEL TRANSFORMATIONS 6-2
       6.2.1   Channel Evolution 6-2
       6.2.2   Channel Incision, Headcuts, and Aggradation 6-3
6.3   STREAM CLASSIFICATION 6-3
       6.3.1   Montgomery and Buffington 6-4
       6.3.2   Rosgen 6-5
       6.3.3   Summary 6-5

TABLE OF CONTENTS (CONTINUED)

Page

CHAPTER 7 - DESIGN PROCEDURE 7-1

7.1 STEP 1. DETERMINE DESIGN FLOWS. 7-3
7.2 STEP 2. DETERMINE PROJECT REACH AND REPRESENTATIVE
CHANNEL CHARACTERISTICS. 7-3
7.3 STEP 3. CHECK FOR DYNAMIC EQUILIBRIUM. 7-4
7.4 STEP 4. ANALYZE AND MITIGATE CHANNEL INSTABILITY. 7-6
7.5 STEP 5. ALIGN AND SIZE CULVERT FOR Q_P. 7-7
    7.5.1 Vertical and Horizontal Alignment. 7-7
    7.5.2 Length 7-8
    7.5.3 Embedment 7-8
    7.5.4 Bed Gradation 7-9
    7.5.5 Manning's n 7-9
    7.5.6 Debris 7-9
    7.5.7 Culvert Analysis and Design Tools 7-10
7.6 STEP 6. CHECK CULVERT BED STABILITY AT Q_H. 7-11
    7.6.1 Permissible Shear Stress 7-11
        7.6.1.1 Noncohesive Materials 7-11
        7.6.1.2 Cohesive Materials 7-13
        7.6.1.3 Applied Shear Stress 7-14
    7.6.2 Critical Unit Discharge 7-14
7.7 STEP 7. CHECK CHANNEL BED MOBILITY AT Q_H. 7-16
7.8 STEP 8. CHECK CULVERT BED STABILITY AT Q_P. 7-16
    7.8.1 Bed Stability 7-17
    7.8.2 Pressure Flow 7-17
7.9 STEP 9. DESIGN STABLE BED FOR Q_P. 7-17
    7.9.1 Oversized Bed Material Gradation 7-18
    7.9.2 Design Equations 7-19
    7.9.3 Design Alternatives 7-19
7.10 STEP 10. CHECK CULVERT VELOCITY AT Q_H. 7-20
7.11 STEP 11. CHECK CULVERT WATER DEPTH AT Q_L. 7-20
7.12 STEP 12. PROVIDE LOW-FLOW CHANNEL IN CULVERT. 7-21
7.13 STEP 13. REVIEW DESIGN. 7-21

CHAPTER 8 - CONSTRUCTION 8-1

8.1 TIMING. 8-1
8.2 STREAM PROTECTION 8-1
8.3 CONSTRUCTABILITY 8-1
8.4 STREAMBED MATERIAL AND PLACMENT 8-1
    8.4.1 Sealing Voids 8-1
    8.4.2 Compaction 8-2

TABLE OF CONTENTS (CONTINUED)

Page

CHAPTER 9 - POST CONSTRUCTION 9-1

9.1   STRUCTURAL INSPECTION... 9-1
9.2   PASSAGE MONITORING 9-1
9.3   MAINTENANCE 9-3

CHAPTER 10 - REFERENCES 10-1

APPENDIX A- METRIC SYSTEM, CONVERSION FACTORS, AND WATER PROPERTIES A-1

APPENDIX B- LEGISLATION AND REGULATION B-1

B.1   FEDERAL STATUTES AND AN EXECUTIVE ORDER..................................... B-1
    B.1.1   Clean Water Act (CWA) 1977 .... B-1
    B.1.2   Endangered Species Act (ESA) of 1973.. B-2
    B.1.3   Fish and Wildlife Coordination Act (FWCA) 1934 .... B-3
    B.1.4   National Environmental Policy Act (NEPA) 1969................................. B-3
    B.1.5   Rivers and Harbors Appropriations Act of 1899 .... B-4
    B.1.6   Sustainable Fisheries Act 1996 .... B-4
    B.1.7   Wild and Scenic Rivers Act 1968.. B-4
    B.1.8   Executive Order on Recreation Fisheries 1995 .... B-4
    B.1.9   Executive Order on Floodplain Management 1977.............................. B-5
B.2   STATE AND LOCAL REGULATIONS B-5

APPENDIX C- MANNING'S ROUGHNESS C-1

C.1   CULVERT MATERIAL... C-1
C.2   BED AND BANK.. C-1
C.3   SAND-BED CHANNELS C-3
C.4   COHESIVE SOILS C-4
C.5   FLOODPLAIN AND VEGETATED BANK ROUGHNESS.................................. C-4
C.6   COMPOSITE ROUGHNESS VALUE C-5

APPENDIX D- PERMISSIBLE SHEAR STRESS D-1

D.1   NONCOHESIVE SOILS D-1
    D.1.1   Shield's Equation... D-1
    D.1.2   Modified Shield's Equation D-2
    D.1.3   Fine-grained Noncohesive Soils D-3

TABLE OF CONTENTS (CONTINUED)

Page

D.2 COHESIVE SOILS D-3
D.3 CRITICAL UNIT DISCHARGE.. D-5

APPENDIX E- EMBEDMENT E-1

APPENDIX F- BED GRADATION F-1

APPENDIX G- BAFFLES AND SILLS G-1

APPENDIX H- DESIGN EXAMPLE: NORTH THOMPSON CREEK, COLORADO ... H-1

H.1 SITE DESCRIPTION.. H-1
H.2 DESIGN PROCEDURE APPLICATION H-3
H.3 SUPPORTING DOCUMENTATION H-17
    H.3.1 Surveyed Cross-sections.. H-17
    H.3.2 HY-8 Report for 6.5 ft CMP at Q_H. H-23
    H.3.3 HEC-RAS Output for 6.5 ft CMP at Q_H. H-27
    H.3.4 HY-8 Report for 8.5 ft CMP at Q_P. H-30
    H.3.5 HEC-RAS Output for 8.5 ft CMP at Q_P. H-34
    H.3.6 HY-8 Report for 8.5 ft CMP with Oversized Bed at Q_H. H-37
    H.3.7 HEC-RAS Output for 8.5 ft CMP with Oversized Bed at Q_H. H-41
H.4 REFERENCES H-1

APPENDIX I- DESIGN EXAMPLE: TRIBUTARY TO BEAR CREEK, ALASKA I-1

I.1 SITE DESCRIPTION.. I-1
I.2 DESIGN PROCEDURE APPLICATION I-1
I.3 SUPPORTING DOCUMENTATION I-12
    I.3.1 Surveyed Cross-sections.. I-13
    I.3.2 HY-8 Report for 7.5 ft CMP at Q_H. I-12
    I.3.3 HEC-RAS Output for 7.5 ft CMP at Q_H. I-20
    I.3.4 HY-8 Report for 12.0 ft CMP with oversized bed at Q_H. I-23
    I.3.5 HEC-RAS Output for 12.0 ft CMP with oversized bed at Q_H............. I-27
I.4 REFERENCES I-30

APPENDIX J- DESIGN EXAMPLE: SICKLE CREEK, MICHIGAN J-1

J.1 SITE DESCRIPTION.. J-1
J.2 DESIGN PROCEDURE APPLICATION J-1

TABLE OF CONTENTS (CONTINUED)

Page

J.3   SUPPORTING DOCUMENTATION J-12

    J.3.1   Cross-sections. J-12

    J.3.2   HY-8 Report for 7.0-ft CMP at Q_H. J-12

    J.3.3   HEC-RAS Output for 7.0-ft CMP at Q_H. J-21

    J.3.4   HY-8 Report for 7.0-ft CMP with Oversize Bed Material at Q_P.......... J-24

    J.3.5   HEC-RAS Output for 7.0-ft CMP with Oversize Bed Material at Q_P.. J-28

J.4   REFERENCES J-31

APPENDIX K- DESIGN EXAMPLE RESULTS COMPARISON K-1

LIST OF TABLES

		Page
Table 3.1	Fish Passage Barrier Types and Their Potential Impacts.	3-2
Table 3.2	Culvert Characteristics for Assessment, Including Possible Barriers.	3-2
Table 4.1	Movement Type as It Relates to Muscle System Utilization.	4-1
Table 4.2	Minimum Depth Criteria for Upstream Passage of Adult Salmon/Trout.	4-4
Table 4.3	Fish Passage Design Criteria for Culvert Installations.	4-5
Table 5.1	State and Agency Guidelines for Q_H.	5-4
Table 5.2	State and Agency Guidelines for Q_L.	5-4
Table 6.1	Stream Classification by Montgomery and Buffington.	6-4
Table 6.2	Morphological Characteristics of the Major Rosgen Stream Types.	6-5
Table 7.1	Selection of Shields' Parameter.	7-12
Table 7.2	Parameter Ranges for Critical Unit Discharge for D_{84}.	7-15
Table 9.1	Monitoring Evaluation.	9-2
Table A.1	Overview of SI.	A-1
Table A.2	Relationship of Mass and Weight.	A-2
Table A.3	Derived Units with Special Names.	A-2
Table A.4	Prefixes.	A-2
Table A.5	Useful Conversion Factors.	A-3
Table A.6	Physical Properties of Water at Atmospheric Pressure (SI).	A-4
Table A.7	Physical Properties of Water at Atmospheric Pressure (CU).	A-5
Table A.8	Sediment Particles Grade Scale.	A-6
Table A.9	Common Equivalent Hydraulic Units: Volume.	A-7
Table A.10	Common Equivalent Hydraulic Units: Rates.	A-7
Table D.1	Selection of Shields' Parameter.	D-2
Table D.2	Coefficients for Permissible Soil Shear Stress.	D-4
Table E.1	Summary of Culvert Embedment Criteria.	E-1
Table H.1	Watershed and Rainfall Characteristics.	H-3
Table H.2	Discharge Estimates.	H-4
Table H.3	Surveyed Cross-Sections.	H-5
Table H.4	Bed Material Quantiles.	H-6
Table H.5	Inlet and Outlet Elevations for Existing and Replacement Culverts.	H-8
Table H.6	Bed Gradation Design.	H-9
Table H.7	Manning's n for Bed Material (D_{84} = 0.56 ft).	H-9
Table H.8	6.5 ft Culvert Inlet and Outlet Parameters at Q_H.	H-12
Table H.9	Estimated Shear Stresses at Q_H.	H-12
Table H.10	7.5 ft Culvert Inlet and Outlet Parameters at Q_H.	H-13
Table H.11	7.5 ft CMP Culvert Inlet and Outlet Parameters at Q_P.	H-14
Table H.12	8.5 ft CMP Culvert Inlet and Outlet Parameters at Q_P.	H-15
Table H.13	Oversize Stable Bed Design Gradation.	H-15
Table H.14	Velocity Estimates at Q_H.	H-16
Table H.15	Maximum Depth Estimates at Q_L.	H-17
Table I.1	Watershed and Rainfall Characteristics.	I-2
Table I.2	Discharge Estimates.	I-2
Table I.3	Surveyed Cross-Sections.	I-3

		Page
Table I.4	Bed Material Quantiles.	I-5
Table I.5	Inlet and Outlet Elevations for Existing and Replacement Culverts.	I-7
Table I.6	Bed Gradation Design.	I-7
Table I.7	Manning's n for Bed Material using the Limerinos Equation.	I-7
Table I.8	7.5 ft CMP Culvert Inlet and Outlet Parameters at Q_H.	I-9
Table I.9	Estimated Unit Discharges at Q_H.	I-9
Table I.10	Oversize Stable Bed Design Gradation.	I-10
Table I.11	Velocity Estimates at Q_H.	I-11
Table I.12	Maximum Depth Estimates at Q_L.	I-11
Table J.1	Watershed and Rainfall Characteristics.	J-2
Table J.2	Discharge Estimates.	J-2
Table J.3	Composited Cross-sections.	J-3
Table J.4	Average Bed Material Quantiles.	J-5
Table J.5	Inlet and Outlet Elevations for Existing and Replacement Culverts.	J-6
Table J.6	Bed Gradation Design.	J-7
Table J.7	Manning's n for Sand Bed Channel.	J-7
Table J.8	7.0-ft CMP Culvert Inlet and Outlet Parameters at Q_H.	J-8
Table J.9	Estimated Shear Stresses at Q_H.	J-9
Table J.10	10.0 ft CMP Culvert Inlet and Outlet Parameters at Q_P.	J-10
Table J.11	Oversize Stable Bed Design Gradation.	J-11
Table J.12	Velocity Estimates at Q_H.	J-11
Table J.13	Maximum Depth Estimates at Q_L.	J-12
Table K.1	Structure Comparisons for Three Case Studies.	K-1

LIST OF FIGURES

Page

Figure 2.1   Changes in Fish Habitat Use Over Time after Roadway Fragmentation. 2-1
Figure 2.2   Perched Outlet, Leap Barrier. 2-2
Figure 2.3   Drop and Velocity Barrier. 2-3
Figure 3.1   Longitudinal Profile Survey Points. 3-3
Figure 4.1   Relative Swimming Abilities of Adult Fish. 4-2
Figure 4.2   Relative Swimming Abilities of Young Fish. 4-3
Figure 4.3   Minimum Water Depths for Fish Passage in Alaska. 4-4
Figure 5.1   Peak Spawning Periods for a Selection of Freshwater Fish. 5-1
Figure 5.2   Example Log-Probability Plot. 5-3
Figure 5.3   Flow Duration Curve for an Annual and a Seasonal Time Period. 5-6
Figure 6.1   Bankfull and Active Channel Widths. 6-1
Figure 6.2   Channel Evolution Model. 6-3
Figure 7.1   Design Procedure Overview. 7-2
Figure C.1   Sand-bed Channel Flow Regime. C-4
Figure C.2   Alternative Estimates for Composite n Values. C-7
Figure D.1   Cohesive Soil Permissible Shear Stress. D-4
Figure H.1   North Thompson Creek H-1
Figure H.2a   North Thompson Creek Culvert Inlet. H-2
Figure H.2b   North Thompson Creek Culvert Outlet. H-2
Figure H.3   North Thompson Creek Drainage Area Delineation. H-3
Figure H.4   Creek and Cross-section Schematic. H-5
Figure H.5   Longitudinal Profile. H-6
Figure H.6   Bed Material Gradation. H-7
Figure I.1   Tributary to Bear Creek Culvert Outlet. I-1
Figure I.2   Creek and Cross-section Schematic. I-4
Figure I.3   Longitudinal Profile. I-4
Figure I.4   Bed Material Gradation. I-5
Figure J.1   Sickle Creek Outlet. J-1
Figure J.2   Creek and Cross-section Schematic. J-4
Figure J.3   Longitudinal Profile. J-4
Figure J.4   Measured Bed Material Gradations. J-5

This page intentionally left blank.

GLOSSARY

Active channel: A waterway of perceptible extent that periodically or continuously contains moving water. It has definite bed and banks, which serve to confine the water and includes stream channels, secondary channels, and braided channels. It is often determined by the "ordinary high water mark" which means that line on the shore established by the fluctuations of water and indicated by a clear natural line impressed on the bank, shelving, changes in the character of soil, changes in vegetation, the presence of litter and debris, or other markers.

Aggradation: The geologic process by which a streambed is raised in elevation by the deposit of material transported from upstream. (Opposite of degradation.)

Apron: A flat or slightly inclined slab up- or downstream of culvert or weir that provides for erosion protection. A downstream apron may also produce hydraulic characteristics that exclude fish.

Anadromous fish: Fish that mature and spend much of their adult life in the ocean, returning to inland waters to spawn. Examples include salmon and steelhead.

Aquatic Organism: Animal growing in, living in, or frequenting water.

Armor: A surficial layer of course grained sediments, usually gravel or coarser, that are underlain by finer grained sediments.

Backwater: Water backed-up or retarded in its course as compared with its normal open channel flow condition. Water level is a function of some downstream hydraulic control.

Baffle: Wood, plastic, concrete or metal mounted in a series on the floor and/or wall of a culvert to increase boundary roughness and thereby reduce the average water velocity in the culvert.

Bed: The bottom of a channel bounded by banks. Also refers to the material placed within an embedded culvert.

Bedform: Elements of the stream channel that describe channel form (e.g. pools, riffles, steps, particle clusters).

Bedload: The part of sediment transport not in suspension consisting of coarse material moving on or near the channel bed.

Bed roughness: Irregularity of streambed material that contributes resistance to streamflow. Commonly characterized using Manning's roughness coefficient.

Bridge: A crossing structure with a combined span (width) greater than 20 ft (6.1 m).

Burst speed: See "Swimming speed."

Cascade: Tumbling flow with continuous jet-and-wake flow over and around individual large rocks or other obstructions. Cascades may be natural or constructed.

Channel: A natural or constructed waterway that has definite bed and banks that confine water.

Channel bed slope: Vertical change with respect to horizontal distance within the channel (Gradient).

Channel bed width: The distance from the bottom of the left bank to the bottom of the right bank. The distinction between bed and bank are determined by examining channel geometry and the presence/absence of vegetation.

GLOSSARY (CONTINUED)

Channelization: Waterway straightening or diverting a waterway into a new channel.

Countersink: Place (embed) culvert invert below stream grade.

Critical depth: The unique depth of flow in a channel that is characteristic only of discharge and channel shape.

Culvert: A conduit or passageway under a road, trail or other waterway obstruction. A culvert differs from a bridge in that it usually consists of structural material around its entire perimeter. A culvert that has a total span (width) of greater than 20 ft (6.1 m) is considered a bridge for purposes of the National Bridge Inspection Standards.

Debris: Includes trees and other organic and inorganic detritus scattered about or accumulated near a culvert by either natural processes or human influences.

Degradation: Erosional removal of streambed material that results in a lowering of the bed elevation throughout a reach. (Opposite of aggradation.)

Deposition: Settlement of material onto the channel bed.

Discharge: Volume of water passing through a channel or conduit per unit time.

> **Bankfull discharge:** Discharge that fills a channel to the point of overflowing onto the floodplain. Generally presumes the channel is in equilibrium and not incising.

> **Channel-forming discharge:** Discharge of water of sufficient magnitude and frequency to have a dominating effect in determining the characteristics and size of the stream course, channel, and bed.

> **Dominant discharge:** Same as channel-forming discharge.

> **Effective discharge:** Discharge that, because of its magnitude and frequency, is responsible for the greatest volume of sediment transport.

Dynamic equilibrium: A stream channel is considered to be in dynamic equilibrium when channel dimensions, slope, and planform do not change radically even though they constantly adjust to changing inputs of water, sediment, and debris.

Embedded culvert: A culvert installation that is countersunk below the stream grade. It may or may not be filled with natural sediment or a design mix.

Entrainment: The process of sediment particle lifting by an agent of erosion.

Entrenchment: The vertical containment of a river and the degree to which it is incised in the valley floor.

Fishway: A system that may include special attraction devices, entrances, collection and transportation channels, a fish ladder, exit and operation and maintenance standards to facilitate passage through bridges or culverts.

Fishway weir: A term frequently used to describe the partition between adjacent pools in a fishway.

GLOSSARY (CONTINUED)

Flood frequency: The frequency with which a flood of a given discharge has the probability of recurring. For example, a "100-year" frequency flood refers to a flood discharge of a magnitude likely to occur on the average of once every 100 years over a very long time span or, more properly, has a 1 percent chance of being exceeded in any year. Although calculation of possible recurrence is often based on historical records, there is no guarantee that a "100-year" flood will occur at all within the 100-year period or that it will not occur several times.

Floodplain: The area adjacent to the stream constructed by the river in the present climate and inundated during periods of high flow.

Flow duration curve: A statistical summary of river flow information over a period of time that describes cumulative percent of time for which flow exceeds specific levels (exceedance flows), exhibited by a cumulative frequency curve that shows the percentage of time that specified discharges are equaled or exceeded. Flow duration curves are usually based on daily streamflow and describe the flow characteristics of a stream throughout a range of discharges without regard to the sequence of occurrence.

Fork length: The length of a fish measured from the most anterior part of the head to the deepest point of the notch in the tail fin.

Geomorphology: The study of physical features associated with landscapes and their evolution. Includes factors such as stream gradient, elevation, parent material, stream size, and valley bottom width.

Grade stabilization or Grade control: Stabilization of the streambed elevation against degradation. Usually a natural or constructed hard point in the channel that maintains a set elevation.

Head-cutting: Channel bottom erosion moving upstream through a stream channel, which may indicate a readjustment of the stream's flow regime (slope, hydraulic control, and/or sediment load characteristics).

Headwater: The water upstream from a structure or point on a stream.

Headwater depth: The depth of water at the inlet of a culvert.

High passage design flow: The maximum discharge used for fish passage design.

Hydraulic jump: Hydraulic phenomenon in open channel flow where supercritical flow changes to sub-critical flow. This results in an abrupt rise in the water surface elevation.

Incision: The resulting change in channel cross-section from the process of degradation.

Interstitial flow: That portion of the surface water that infiltrates the streambed and moves through the substrate interstitial spaces.

Invert: The lowest point of the internal cross section of culvert.

Large Woody Debris (LWD): Any large piece of woody material such as root wads, logs and trees that intrude into a stream channel. LWD may occur naturally or be designed as part of a stream restoration project.

Low passage design flow: The minimum discharge used in fish passage design.

Manning's n: Empirical coefficient for simulating the effect of wetted perimeter roughness used in determining water velocity in stream discharge calculations.

GLOSSARY (CONTINUED)

Mitigation: Actions to avoid or compensate for the impacts resulting from a proposed activity.

Normal depth: The depth of flow in a channel or culvert when the slopes of the water surface and channel bottom are the same.

Perching: The tendency to develop a scour hole at the outfall of a culvert due to erosion of the stream channel.

Pipe: A culvert that is circular (round) in cross section.

Pipe arch: A pipe that has been factory-deformed from a circular shape such that the span (width) is larger than the vertical dimension (rise).

Plunging flow: Flow over a weir or out of a perched culvert, which falls into a receiving pool.

Regrade: The process of channel adjustment to attain a new "stable" bed slope. For example, following channelization, a streambed will typically steepen upstream and flatten downstream.

Resident fish: Fish that migrate and complete their life cycle in fresh water.

Riparian: The area adjacent to flowing water (e.g., rivers, perennial or intermittent streams, seeps or springs) that contains elements of both aquatic and terrestrial ecosystems that mutually influence each other.

Riprap: Large, durable materials (usually rocks; sometimes broken concrete, etc.) used to protect a stream bank from erosion and other applications.

Scour: Localized erosion caused by flowing water.

Shear stress: Hydraulic stress (force per unit area) of water created by its movement across a submerged surface such as the channel bed or channel bank.

Substrate: Mineral and organic material that forms the bed of a stream. In an armored channel, substrate refers to the material beneath the armor layer.

Supercritical flow: Occurs when normal depth is less than critical depth; rare for extended reaches in natural streams.

Swimming speeds: Fish swimming speeds can vary from essentially zero to over six meters per second, depending on species, size and activity. Three categories of performance are generally recognized based on the duration of swimming to when a fish becomes fatigued and requires rest:

> **Sustained speed:** The speed a fish can maintain for an extended period for travel without fatigue. Metabolic activity in this mode is strictly aerobic and utilizes only red muscle tissues.

> **Prolonged speed:** The speed that a fish can maintain for a prolonged period, but which ultimately results in fatigue. Metabolic activity in this mode is both anaerobic and aerobic and utilizes white and red muscle tissue.

> **Burst (Darting) speed:** The speed a fish can maintain for a very short period, generally 5 to 7 seconds, but less than 15 seconds, without gross variation in performance. A rest period is required. Burst speed is employed for feeding, escape and negotiating difficult hydraulic situations, and represents maximum swimming speed. Metabolic activity in this mode is strictly anaerobic and utilizes only white muscle tissue.

GLOSSARY (CONTINUED)

Tailwater: The water downstream from a structure or point on a stream.

Tailwater depth: Depth of water at a culvert outlet.

Thalweg: The longitudinal line of deepest water within a stream.

Toe: The break in slope at the foot of a bank where the stream bank meets the bed.

Upstream passage facility: A fishway system designed to pass fish upstream of a passage impediment, either by volitional or non-volitional passage.

Velocity: Time rate of motion; the distance traveled divided by the time required to travel that distance.

> **Average velocity:** The discharge divided by the cross-sectional area of the flow in a culvert or channel cross-section.

> **Boundary layer velocity:** Area of decreased velocity resulting from boundary roughness. This region is restricted to only a few centimeters from the boundary.

> **Maximum velocity:** The highest velocity within a cross-section of flow.

Weir: A short wall constructed on a stream channel that backs up water behind it and allows flow over or through it if notched. Weirs are used to control water depth and velocity.

Wetted perimeter: The boundary over which water flows in a channel, stream, river, swale, or drainage facility such as a culvert or storm drain.

This page left intentionally blank.

GLOSSARY OF ACRONYMS

Acronym	Definition
ADFG	Alaska Department of Fish and Game
ADOT	Alaska Department of Transportation
AOP	Aquatic Organism Passage
BMP	Best Management Practices
CALTRANS	California Department of Transportation
CDFG	California Department of Fish and Game
CMP	Corrugated Metal Pipe
CWA	Clean Water Act
CU	Customary Units
DF&G	Department of Fish and Game
DOT(s)	Department(s) of Transportation
EDF	Energy Dissipation Factor
EFH	Essential Fish Habitat
EO	Executive Order
ESA	Endangered Species Act
FDC	Flow Duration Curves
FHWA	Federal Highway Administration
FWCA	Fish and Wildlife Coordination Act
FSSWG	Forest Service Stream-Simulation Working Group
GAO	General Accounting Office
HDS	Hydraulic Design Series
HEC	Hydraulic Engineering Circular
MDOT	Maine Department of Transportation
NEPA	National Environmental Policy Act
NMFS	National Marine Fisheries Service
NOAA	National Oceanic-Atmospheric Administration
ODFW	Oregon Department of Fish and Wildlife
OHW	Ordinary High Water
QA/QC	Quality Assurance and Quality Control
SI	International System of Units
SPP	Structural Plate Pipe
SPPA	Structural Plate Pipe Arch
WDFW	Washington Department of Fish and Wildlife
USFS	United States Forest Service

This page left intentionally blank.

LIST OF SYMBOLS

Symbol	Definition
A	area, ft^2 (m^2)
b_c	channel span across bars, ft (m)
C_d	discharge coefficient
C_e	dimensionless culvert exit head loss coefficient
C_0	dimensionless culvert head loss coefficient (C_e+K_e)
d	particle size of interest, ft (m)
D_i	particle size representing i percent finer
	(Example, D_{16} is the particle size representing 16 percent finer)
f	dimensionless Darcy Weisbach friction factor
F*	dimensionless Shield's parameter
g	acceleration due to gravity, ft/s^2 (m/s^2)
h	bank height, ft (m)
HW	headwater depth above the culvert entrance invert, ft (m)
K_e	dimensionless culvert entrance head loss coefficient
L	length, ft (m)
h_t	critical bank height, ft (m)
m	Fuller-Thompson parameter for adjusting bed mixture gradation
n	Manning's roughness coefficient
Q	flow, ft^3/s (m^3/s)
q	unit discharge, $ft^3/s/ft$ $(m^3/s/m)$
q_c	critical unit discharge, $ft^3/s/ft$ $(m^3/s/m)$
Q_{100}	one hundred year flow, ft^3/s (m^3/s)
R	hydraulic radius, ft (m)
S	slope, ft/ft (m/m)
S_f	friction (energy) slope, ft/ft (m/m)
V	velocity, ft/s (m/s)
y	depth of water, ft (m)
Z	baffle height, ft (m)
τ	shear stress, lb/ft^2 (N/m^2)
τ_c	critical shear stress, lb/ft^2 (N/m^2)
γ	specific weight of water, lb/ft^3 (N/m^3)
Φ	angle of repose, degrees (radians)

This page intentionally left blank.

CHAPTER 1 - INTRODUCTION

1.1 PURPOSE

This manual presents a stream simulation design procedure, methods and best practices for designing culverts to facilitate aquatic organism passage (AOP). Although this manual focuses on culverts, the design team should recognize that an appropriate structure for any given crossing may be a bridge. This manual is not intended to conflict with or replace accepted guidance and procedures adopted in particular locations. When specific water crossing design methods are required in the jurisdiction where the crossing is located, those methods should be applied. In addition, local and regional requirements may overlay additional steps on this design approach.

Since fish have been the primary focus of AOP design efforts over the years, and much has been learned about fish specifically, many of the references to AOP in this manual derive directly from what is known about fish. However, the broader scope of AOP is the focus of the manual.

Because of the variety of fish and other aquatic species in the U.S., the complex nature of fish behavior, and the variation in such behaviors and capabilities over the various life-stages, designing hydraulic structures with satisfactory aquatic organism passage (AOP) characteristics remains a challenging endeavor. Over the years, resource agencies and others have assembled a large amount of empirical data and field experience to guide the design of roadway structures, particularly culverts, for passage. Much of the resulting criteria are based upon the natural geomorphic characteristics of streams supporting the aquatic ecosystems of interest, and many of the procedures implementing those criteria seek to replicate the stream and floodplain characteristics and geometries within the roadway crossing structure. The "stream simulation" approach such as developed by the United States Forest Service (FSSWG, 2008) is one approach that is state of the art.

Given the diverse behavior and capabilities of fish and other aquatic organisms, design procedures necessarily rely on surrogate parameters and indicators as measures for successful passage design. Many of the existing AOP design procedures rely on dimensional characteristics of the stream such as bankfull width. A critique of the use of dimensional stream characteristics is that they: 1) can be difficult to identify, 2) can be highly variable within a stream reach, 3) assume the stream is in dynamic equilibrium, and 4) have no known relationship to passage requirements.

The procedure described in this manual uses streambed sediment behavior as its surrogate parameter. The hypothesis of using sediment behavior as a surrogate parameter is that aquatic organisms in the stream are exposed to similar forces and stresses experienced by the streambed material. The design goal is to provide a stream crossing that has an equivalent effect, over a range of stream flows, on the streambed material within the culvert compared with the streambed material upstream and downstream of the culvert. When this is achieved and the velocities and depths are comparable to those occurring in the stream, the conditions through the crossing should present no more of an obstacle to aquatic organisms than conditions in the adjacent natural channel.

The primary goal of this document is to incorporate many of the current geomorphic-based design approaches for AOP while providing a procedure based on quantitative best practices. The stream simulation design procedure is intended to create conditions within the crossing similar to those conditions in the natural channel to provide for aquatic organism passage (AOP). This document seeks to identify, develop, and present a bed stability-based approach that accounts for the physical processes related to the natural hydraulic, stream stability, and sediment transport characteristics of a particular stream crossing as surrogate measures.

1.2 CONTEXT

1.2.1 Historical Crossing Design

Waterway crossings, including bridges and culverts, represent a key element in our overall transportation system. The design of crossing structures has traditionally used hydraulic conveyance and flood capacity as the main design parameters. Hydraulic Design Series - 5 *Hydraulic Design of Highway Culverts* (HDS 5) specifies a culvert design procedure to maintain acceptable headwater depth during design floods; this ensures efficient conveyance of water, but normally does not include provisions for aquatic organism passage (AOP) through the culvert (Normann, et al., 2005).

Crossing structures often narrow the channel through the bridge opening or culvert barrel. Constricted reaches influence the characteristics of flow through and around the hydraulic structure, increasing velocities and scour potential (Johnson and Brown, 2000). High flow regimes may induce scour of the streambed through and downstream from the structure, and cause upstream progressing channel incision (Castro, 2003).

1.2.2 Road Stream Interaction

Roads cover almost two percent of the landmass in the United States, leading to a seemingly unavoidable interaction of roadways and the environment (Schrag, 2003). For example, a survey of Bureau of Land Management (BLM) and U.S. Forest Service land found 10,000 culvert crossings on fish bearing streams in Washington and Oregon with over half considered to be barriers to juvenile salmon passage (General Accounting Office, 2001). Estimates of road and railroad crossing affecting Massachusetts streams are as high as 28,500 (Venner Consulting and Parsons Brinkerhoff, 2004). Such crossings impact aquatic organisms, including fish, potentially causing barriers to passage, fragmentation, and a loss of ecological connectivity (Trombulak and Frissell, 2000). Many of the culverts that are currently in place were designed and installed with hydraulic conveyance as the main criterion (Normann et al., 2005). Natural stream processes and aquatic organism passage (AOP) were generally not considered in designing culverts that could pass a design flow without roadway overtopping.

Although much focus has been on the passage of fish, many other organisms are affected by culverts designed without passage consideration, including small aquatic organisms such as salamanders (United States Forest Service, 2006a; Schrag, 2003). In general, a culvert that is impassable for fish may also pose a barrier to aquatic organisms including those with weaker swimming abilities (FSSWG, 2008).

As increasing human population leads to an expansion of our infrastructure, the role of roads in habitat decline and fragmentation is the subject of increased scrutiny (e.g. Spellerberg, 1998; Trombulak and Frissell, 2000). The long-term ecological effects of roads include loss and change of habitat, changes in biological makeup of communities, and fragmentation – leading to population isolation (Spellerberg, 1998).

River and stream corridors provide vital habitat for a wide range of animal species, many of which depend on the ability to move freely throughout their ecosystem in order to complete their life cycles (Jackson, 2003). The importance of human transportation has led to roads that extend through much of the country, inevitably crossing over streams and rivers. Frequently, the design of structures to pass water under a road did not consider animal movement, causing fragmentation of many riverine systems (Trombulak and Frissell, 2000). Recognition of the need to restore habitat connectivity has added ecological considerations to the design and retrofit of road stream crossings (e.g. Jackson, 2003; FSSWG, 2008).

1.3 DESIGN PROCEDURE APPLICABILITY

The design procedure featured in this document is applicable nationwide and targets embedded culverts. An embedded culvert is a closed-bottom conduit with the bottom buried (embedded) a certain depth. The procedure applies to single or multiple barrel culvert installations. However, multiple barrel installations may not be preferred in some situations because they divide the flow.

The procedure may also be applied to open-bottom culverts. There are two primary differences between open-bottom culverts and embedded closed-bottom culverts. First, an open-bottom culvert does not explicitly provide a grade control function. However, grade control may be installed upstream, downstream, or within an open-bottom culvert. Second, with an open-bottom culvert there is no need to specify an embedment depth, though scour must be considered in designing the foundation depth. Considerations such as construction dewatering, site geology, span, and cost, as well as the results of the application of this design procedure, should be considered in choosing between an open-bottom or embedded culvert.

The procedure is applicable to new crossing sites as well as culvert replacements. As discussed in Chapter 7, the designer must be aware of the impacts of the existing culvert on the stream as part of planning and designing for the replacement.

This procedure does not target specific organisms or life stages for passage, nor does the designer need to match species-specific water velocity, water depth, or crossing length criteria as is necessary for some design procedures. Specific information on fish or other aquatic organisms is not required to successfully apply the design procedure. However, consultation with appropriate state and Federal agencies may result in the identification of specific species of concern at a given site. If this is the case, any additional criteria related to those species should be incorporated into the design through coordination with the appropriate agencies.

1.4 MANUAL ORGANIZATION

The next five chapters provide background information a designer should be familiar with including how culverts create passage barriers, techniques for AOP culvert assessments and inventories, fish biology, fish passage hydrology, and stream geomorphology. The design procedure is described in Chapter 7. The final two chapters cover issues of construction and post-construction. Detailed technical information supporting the methods used within the design procedure and several design examples are included in the appendices. These best practices should be evaluated as research and application experience advances.

As will become readily apparent from reading this manual and applying the design procedure, AOP design, construction, and monitoring is a multi-disciplinary activity that often requires a team that includes several of the following disciplines: aquatic biology, geomorphology, hydrology, sediment transport, hydraulic engineering, and geotechnical engineering. It is also critically important for the design team to coordinate early and often with local, regional, state, and Federal permitting authorities.

This page intentionally left blank.

CHAPTER 2 - CULVERTS AS PASSAGE BARRIERS

2.1 STREAM FRAGMENTATION

Culvert installations can significantly decrease the probability of aquatic organism movement between habitat patches (Schaefer et al. 2003). Figure 2.1 depicts the possible results of ineffective culverts on fish populations. In the undisturbed case, fish are free to use the entire stream system as habitat. After a road interrupts stream continuity, fragmented populations are forced to survive independently. In a short time frame, this interruption in continuity increases the susceptibility of smaller populations to elimination by chance events (Farhig and Merriam, 1985). Over the long-term, genetic homogeneity and natural disturbances are also likely to destroy larger populations (Jackson 2003). Figure 2.1 shows this process sequentially from top left to bottom right: (a) undisturbed habitat, with fill representing habitat in use; (b) habitat with ineffective culverts causing fragmentation with fill colors representing disconnected habitats; (c) fragmented system after a few years, areas with no fill represent population extirpation; (d) fragmented system after many years.

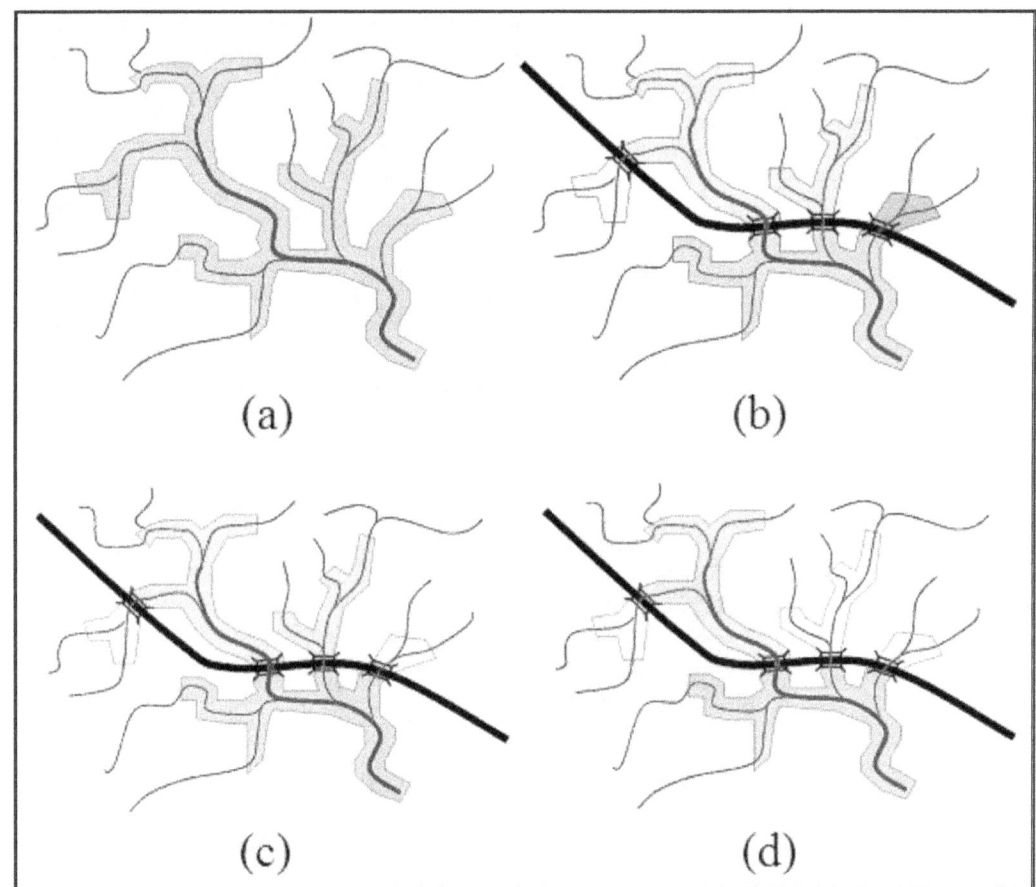

Figure 2.1.   Changes in Fish Habitat Use Over Time after Roadway Fragmentation.

2.2 BARRIER MECHANISMS

A culvert becomes a barrier to AOP when it poses conditions that exceed the organism's physical capabilities. Circumstances that serve as barriers are species-dependent; therefore, the balance of this section addresses barriers to fish passage. Common obstructions include excessive water velocities, drops at culvert inlets or outlets, physical barriers such as weirs, baffles, or debris caught in the culvert barrel, excessive turbulence caused by inlet contraction, and low flows that provide too little depth for fish to swim.

The severity of obstacles to passage intensifies when a series of obstacles cause fish to reach exhaustion before successfully navigating the structure. For example, fish have been observed successfully passing an outlet drop, but having insufficient white muscle capacity to traverse a drop upon reaching the culvert inlet (Behlke et al., 1989). As noted in Chapter 4, fish swimming abilities are not cumulative, and a fish that reaches exhaustion in any category of muscle use will require a period of rest before continued movement (Bell, 1986).

2.2.1 Drop at Culvert Outlet

Drops in water surface will create passage barriers when they exceed fish jumping ability. Drops can occur at any contiguous surface within the culvert, but they are most commonly seen at the culvert outlet (see Figure 2.2 provided by the Alaska Department of Fish and Game (2005)), where scour and downstream erosion leads to culvert perching (Forest Practices Advisory Committee on Salmon in Watersheds, 2001). See Chapter 4 for examples of species-specific jumping abilities.

Figure 2.2.   Perched Outlet, Leap Barrier.

2.2.2 Outlet Pool Depth

Fish that will jump require a jump-pool to gain the momentum necessary to jump into the structure. Early field observations of salmon and trout suggested that successful passage at falls occurs when the ratio of the drop height to pool depth is greater than or equal to 1.25 (Stuart 1962). Aaserude and Orsborn (1985) later correlated fish passage to fish length and the depth that water from the falls penetrates the pool. For practical application, jump pool requirements are generally specified based on a ratio of pool depth to drop height. Oregon, for example, uses 1.5 times jump height, or a minimum of 2 ft (0.6 m), for pool depth (Robison, et al., 1999). However, an adequate jump-pool does not guarantee that a fish has the ability to make the required leap, or once in the culvert, has the energy to overcome the water velocity in the culvert barrel.

An additional factor in the pool depth assessment is the size of the fish related to the size/depth of the pool. For a given pool size, a larger fish may have more difficulty with a pool than a smaller fish because it may have insufficient space to initiate and execute a jump.

2.2.3 Excessive Barrel Velocity

Figure 2.3 (Alaska Department of Fish and Game, 2005) depicts a culvert outlet presenting a drop and velocity barrier to fish passage. There are many categories of velocity that affect fish passage within a culvert crossing. These include boundary layer velocity, maximum point velocity, average cross-sectional velocity, and inlet transition velocity. The importance of each is discussed below.

Figure 2.3. Drop and Velocity Barrier.

2.2.3.1 Boundary Layer Velocity

Due to the no-slip condition in fluid mechanics, water velocity at all points of contact with the culvert is zero. The velocity increases away from the boundary, forming a so-called boundary layer. Boundary roughness increases the depth of reduced velocity. Fish have been observed to use this area to hold and rest, or swim upstream through culverts (Behlke et al., 1989; Powers et al. 1997). Investigation of the development of low velocity zones has quantified velocity reduction in round culverts for use in fish passage design (Barber and Downs, 1996). However, variability in flow patterns and fish utilization is likely too great for this phenomenon to be consistently accounted for in design standards (Lang et al., 2004). To ensure passage, Powers, et al. (1997) recommended that design be based on average cross-sectional velocity - without direct considerations of roughness. Although the impacts of roughness have not been directly correlated to fish passage success in the field, using corrugated pipe and large corrugations is still common practice to increase roughness and decrease boundary layer velocity (e.g. Maine Department of Transportation, 2004; Bates et al., 2003; Robison et al., 1999).

2.2.3.2 Average Velocity

Average cross-sectional velocity is the most common velocity parameter used in culvert design. Although the characteristics of a fish's chosen path may not be well represented by average velocity (Powers et al., 1997; Barber and Downs, 1996), little is understood about the utilization and development of boundary layers within a culvert, and average velocity represents a conservative design parameter (Lang et al., 2004).

2.2.3.3 Maximum Point Velocity

Points of maximum velocity will also occur within the culvert as water flows over or around constrictions such as weirs or baffles. While average design velocity will more likely be relevant to a fish's prolonged swimming ability, fish may be required to use their white muscle tissue to burst through zones of maximum velocity (Rajaratnam et al., 1991).

2.2.3.4 Inlet Transition Velocity

The culvert inlet requires special consideration, as it is the last barrier for a fish traversing a culvert. Velocity at the inlet may be higher than in the barrel if bedload deposits upstream from the entrance increase the local slope. Inlet conditions are especially important in long installations, or when successful navigation through a series of other obstacles has required significant use of fishes' white muscle tissue. The addition of tapered wingwalls may significantly reduce the severity of an inlet transition (Behlke et al., 1991). A skewed entrance will produce higher entrance velocities than a non-skewed entrance.

2.2.4 Insufficient Depth

Insufficient depth can be a barrier within the culvert or on any continuous flow area upstream or downstream of the culvert installation. Insufficient depth will impair fishes' ability to generate maximum thrust, increase fishes' contact with the channel bottom, and reduce the fishes' ability to gather oxygen from the water (Dane, 1978). Combined, these effects reduce a fish's swimming potential and increase the risk of bodily injury and predation. Sufficient depth is also required to support the fish while resting.

2.2.5 Excessive Turbulence

Treatments used to reduce culvert velocity or increase depth may also increase turbulence, and dissuade fish from entering or traversing the structure or confuse their sense of direction. Although little is understood about the effects of turbulence on fish passage, recent studies at the University of Idaho have found that fish prefer to hold in zones of low turbulence (Smith and Brannon, 2006). Washington DOT and Maine DOT design guidelines suggest fish turbulence thresholds, quantifying turbulence with an Energy Dissipation Factor (EDF) (Bates et al., 2003; Maine Department of Transportation, 2004):

$$EDF = \gamma QS / A \qquad\qquad (2.1)$$

where,

$\quad$ EDF $\;=\;$ Energy Dissipation Factor, ft-lb/ft^3/s (m-N/m^3/s)

$\quad \gamma \quad=\;$ unit weight of water, lb/ft^3 (N/m^3)

$\quad$ Q $\quad=\;$ flow, ft^3/s (m^3/s)

$\quad$ S $\quad=\;$ slope of the culvert, ft/ft (m/m)

$\quad$ A $\quad=\;$ cross-sectional flow area, ft^2 (m^2)

Washington State suggests the EDF be less than 7.0 ft-lb/ft^3/s (335 m-N/m^3/s) for roughened channels, 4.0 ft-lb/ft^3/s (191 m-N/m^3/s) for fishways, and 3.0-5.0 ft-lb/ft^3/s (144-239 m-N/m^3/s) for baffled culvert installations. These criteria are based on experience in Washington, and should be evaluated with future research and experience (Bates et al., 2003). Maine DOT has similar guidelines (Maine Department of Transportation, 2004).

2.2.6 Culvert Length

Longer culvert installations require fish to maintain speed for extended periods of time, leading to increased energy expenditure. For this reason, maximum allowable velocity thresholds decrease with increasing culvert length (Bates et al., 2003; Robison et al., 1999). Longer culverts with natural substrate may not represent a barrier if fish can rest in reduced velocity zones.

Extreme length can also cause a culvert to be dark. Research has noted behavioral differences in light versus dark passage of fish species (Welton et al., 2002; Kemp et al., 2006; Stuart, 1962), suggesting that darkness may dissuade certain fish from entering a structure (Weaver et al. 1976). This theory has yet to be accepted as common knowledge (Gregory et al., 2004), but deserves consideration when installations require long structures. However, there is no quantitative definition of "long" in this context; qualitatively a "long" culvert is one that discourages passage as a direct result of its length.

2.2.7 Debris and Sediment Accumulation

Culverts with baffles, large roughness elements, or small diameters may have a high propensity to collect debris. This debris can include natural materials such as Large Woody Debris (LWD) and warrants specific consideration in areas where anthropogenic or natural debris accumulation is likely. A monitoring and maintenance program can identify culverts that require more attention than others (Forest Practices Advisory Committee on Salmon in Watersheds, 2001). Sediment accumulation at a culvert entrance may also be a barrier to passage.

2.2.8 Culvert Damage

Some culverts may exhibit damage at the entrance or exit, as well as within the barrel. These unfamiliar conditions may dissuade fish from attempting passage. This concern can be avoided by simply maintaining good operating conditions at all culvert installations.

This page intentionally left blank.

CHAPTER 3 - AOP CULVERT ASSESSMENT AND INVENTORY

3.1 AOP CULVERT ASSESSMENT

Procedures and criteria for AOP culvert assessment must be developed to support barrier removal and habitat restoration program goals. Properly designed culvert assessment will provide adequate knowledge of a crossing location and ultimately lead to a robust inventory. Agreements between State DOTs and Resource agencies can greatly expedite the design and assessment procedure, ensuring that the requirements of all parties are met satisfactorily through a common vision. For example, Alaska and Oregon currently have agreements between their respective resource agencies to expedite permit applications with respect to AOP at culvert installations. They also have a shared priority of replacement/repair of fish passage barriers (Venner Consulting and Parsons Brinkerhoff, 2004).

3.1.1 Assessment Criteria

Before crossing assessment can begin, it is necessary to have a clearly defined set of assessment criteria. Much like culvert design criteria, assessment criteria show regional variability, but generally consider the following elements to determine fish passability:

- Flow depth
- Flow velocity
- Drop heights
- Pool depths
- Culvert length
- Culvert type (shape and material)
- Culvert condition
- Culvert orientation
- Substrate
- Site stability
- Aggradation and degradation at culvert inlet and outlet

Assessment criteria are based on fish species present as well as the timing and duration of fish movement. Criteria for adult salmon, for example, will be significantly different from that used for juveniles or trout species (e.g. Robison et al., 1999; Washington Department of Fish and Wildlife, 2000).

It is recommended that assessment criteria be developed separately from design criteria (Lang et al., 2004). Typically, design criteria are conservative, so as to provide passage for the weakest swimming individual during a range of design flows. Assessment criteria, however, seek to determine the degree to which a crossing is a barrier to fish passage. Crossings that would be labeled inadequate by design standards may only provide a partial barrier to fish passage. As a result, criteria for design and assessment are slightly different, and generally not interchangeable.

3.1.2 Degree of Barrier

Assessment allows crossings to be grouped into broad categories of adequacy such as "Passable," "Impassable," and "Indeterminate." Category definitions are expounded to clearly place barriers within a matrix. In California, a culvert that can pass all salmonids during the entire migration period earns a "green" classification, while a culvert that does not meet requirements of strongest swimming fish and life stage present over the entire migration period is classified as "red," analogous to traffic signalization (Taylor and Love, 2003). Culverts that cannot be placed in these categories remain in the "gray" or "indeterminate" area, where the crossing may present impassable conditions to some species and life stages at some flows. Further analysis is required in order to ascertain the extent of the barrier.

It is likely that initial surveys will show many culverts to be "indeterminate," where adequacy cannot be determined without a detailed hydraulic analysis (Clarkin, et al., 2003). Furthermore, a great number of "impassable" crossings typically ensure that "indeterminate" crossings are never properly analyzed (Furniss, 2006).

Culverts falling into the "indeterminate" area are likely to be barriers to some fish species and life stages. Table 3.1 shows barrier categories used in California (Taylor and Love, 2003). Assessment criteria are used to prioritize culvert crossings for future replacement, and the degree of barrier is one of many factors used to determine the urgency of culvert replacement/retrofit. The traditional design approach for culverts has resulted in many that are a partial or temporal barrier to fish passage. An understanding of the degree to which a culvert is a barrier is useful in assessing the effect of that culvert on the surrounding ecosystem and in determining the need and urgency of culvert replacement (Furniss, 2006).

Table 3.1.   Fish Passage Barrier Types and Their Potential Impacts.

Barrier Category	Definition	Potential Impacts
Temporal	Impassable to all fish at certain flow conditions (based on run timing and flow conditions)	Delay in movement beyond the barrier for some period of time
Partial	Impassable to some fish species, during part or all life stages at all flows.	Exclusion of certain species during their life stages from portions of a watershed
Total	Impassable to all fish at all flows	Exclusion of all species from portions of a watershed.

3.1.3 Data Collection

An initial survey of the culvert and adjoining stream reach will allow a basic understanding of stream crossing conditions. This survey should cover a number of site characteristics including culvert and channel measurements and classification, flow data, and watershed conditions. Specific culvert characteristics of interest may include those listed in Table 3.2 (Coffman, 2005). It will be useful to have a standardized survey collection method that incorporates collection of all pertinent parameters.

Table 3.2.   Culvert Characteristics for Assessment, Including Possible Barriers.

Culvert Characteristic	Possible Barrier
Outlet drop and outlet perch	Jump barrier
Culvert slope	Velocity barrier
Culvert slope times length	Exhaustion barrier
Presence of natural stream substrate	Depth barrier
Relationship of tailwater control elevation to culvert inlet elevation	Depth and velocity barrier

Basic survey techniques are included in *Stream Channel Reference Sites: An Illustrated Guide to Field Technique* (Harrelson et al., 1994). Examples of fish passage survey applications, including forms, explanations of survey points, and data collection are included in Appendix E of *National Inventory and Assessment Procedure* (Clarkin et al., 2003). Taken from Clarkin, et al. (2003), Figure 3.1 depicts some typical longitudinal survey points used in a culvert survey.

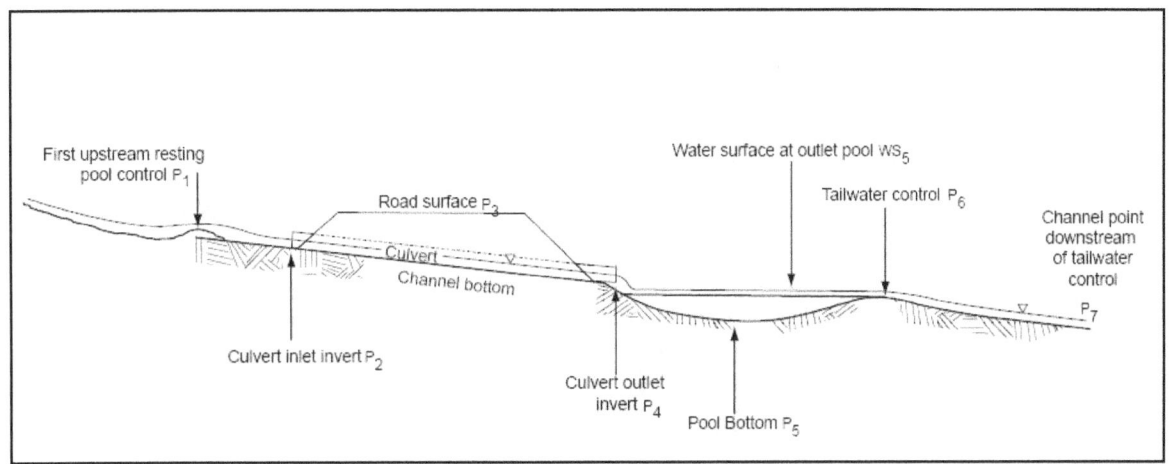

Figure 3.1.   Longitudinal Profile Survey Points.

3.2   CULVERT INVENTORY

The first step in a program of fish passage restoration is awareness of the problem, including location and condition of waterway crossings. An inventory can be as simple as a listing of the locations of existing roadway-stream crossings, and will ideally include basic survey information. A robust inventory will be invaluable in planning efforts and many assessment schemes have been created to collect information necessary for the prioritization of crossing replacement, e.g. Clarkin, et al., 2003; Taylor and Love, 2003; Washington Department of Fish and Wildlife, 2000. There are two standard methods for completing a culvert inventory, including road- and stream-based approaches. Departments of transportation typically use road-based inventories, while stream-based inventories are usually performed by resource agencies.

A national inventory process created by the Forest Service was designed to answer two questions (Clarkin, et al., 2003):

1.   Does the crossing provide adequate passage for the species and life-stage of concern?

2.   What is the approximate cost of replacement?

An inventory allows a basic understanding of fish impediments, as well as the requirements/plausibility of replacement. Additional information, such as environmental risk, may also be beneficial to planners attempting to prioritize corrections of roadway-stream treatments. Risk assessments may be coupled with fish passage assessment and inventories, but will require additional time and expense. Methods for determining environmental risk are outlined in *Methods for Inventory and Environmental Risk Assessment of Road Drainage Crossings* (Flanagan et al., 1998).

3.2.1   Road-based Inventory

A road-based inventory follows a particular road system to identify and evaluate all road stream crossings. This type of inventory is useful to managers requiring knowledge of highway effects on fish passage, and allows highway dollars to be efficiently spent on the mitigation of fish passage barriers. For example, minor adjustment to culvert inlet or outlet conditions, such as debris jams, rock placement, backwatering, etc., can be made during routine road maintenance. Known barriers can be addressed as part of rehabilitation or reconstruction projects.

Road-based approaches can be comprehensive, although following a road will invariably miss a number of barriers that exist on side streams or barriers created by minor roads, manmade dams, or diversions (Washington Department of Fish and Wildlife, 2000).

3.2.2 Stream-based Inventory

A stream-based inventory follows the entire fish bearing channel system within a watershed, noting all constructed obstacles (e.g. dams, culverts, water diversions). Paramount to the inventory is information on the species of concern and their spatial, temporal, and life stage habitat requirements. Further evaluation of these structures provides an understanding of fish passage barriers in a watershed context.

This type of inventory will allow analysis of the extent of stream habitat that can be opened up by repairing/replacing a particular culvert. This information serves as the basis for the biological and economic evaluation of benefits to ensure that program dollars are well spent. Effective inventories and repair/replacement prioritizations often require cooperation amongst the agencies that have jurisdiction along a stream corridor.

CHAPTER 4 - FISH BIOLOGY

The design procedure in this document does not require an assessment of fish biology at a particular site. However, an understanding of fish biology and swimming ability is useful. The following discussion outlines fish biology, swimming abilities, and requirements, providing a basic understanding of what fish need to successfully move throughout their environment.

4.1 CAPABILITIES AND ABILITIES

Fish possess two muscle systems to accommodate different modes of travel: a red muscle system (aerobic) for low-intensity activities and a white muscle system (anaerobic) for shorter, high-intensity movements (Webb, 1975). Extensive use of the white muscle system causes extreme fatigue, requiring extended periods of rest.

4.1.1 Swimming and Jumping

Fish movement can be divided into three categories based on speed and muscle use: sustained, prolonged or burst speeds (Bell, 1986). A fish at sustained speed uses the red muscle system exclusively, allowing extended periods of travel at low speeds. Prolonged speed involves the use of both red and white muscle tissue, and allows the fish to reach quicker speeds for minutes at a time. Burst speed allows the fish to reach top speeds for a few seconds by exclusive utilization of white muscle tissue, requiring a significant rest period. Table 4.1 (adapted from Bell, 1986 and Powers and Orsborn, 1985)) summarizes the muscle system use as it relates to fish movement.

Table 4.1.   Movement Type as It Relates to Muscle System Utilization.

Movement Type	Description	Muscle System	Period
Sustained	Used for long periods of travel at low speeds. Normal functions without fatigue.	Red (purely aerobic)	Hours or days
Prolonged	Short periods of travel at high speeds resulting in fatigue	Red and White	0.25 to 200 minutes
Burst	Maximum swimming speed or jumping, inducing fatigue.	White (purely anaerobic)	0 to 15 seconds

Fish can fail to pass a culvert for a variety of reasons. An outlet drop or high velocity zone will act as a barrier when it exceeds the fish's burst swimming ability, while a long continuous section of culvert with relatively low velocity may require prolonged swimming speeds to be maintained beyond a fish's natural ability. It is important to note that these criteria are not cumulative, and a fish that reaches exhaustion in any category will require a period of rest before continued movement.

A number of studies have been completed to ascertain the swimming and jumping ability of different fish species (e.g. Jones et al., 1974; Bainbridge, 1959; Stuart, 1962; Hinch and Rand, 1998; Rand and Hinch, 1998; Ellis 1974; Toepfer et al. 1999). An excellent database is maintained within the US Forest Service FishXing computer program (US Forest Service, 2006b).

Design to meet the needs of a spawning salmon will not necessarily guarantee that a culvert will allow passage of weaker swimming juveniles or resident fish. Although fish are capable of specific swimming energies, it does not mean that fish will choose to expend maximum swimming energy when confronted with specific obstacles (Behlke et al. 1991). This is consistent with observations of fish moving through culvert boundary layers, and holding in areas of low velocity between corrugations (Powers et al. 1997).

4.1.2 Species and Life Stages

Swimming and jumping capabilities can vary greatly between species. A significant portion of the variability is related to body mass, that is, the greater the body mass, the greater the capability. For example, Figure 4.1, taken from Bell's Fisheries Handbook (1986), depicts the relative swimming abilities of adult fish. Burst speeds reaching 26 ft/s (7.9 m/s) give adult steelhead a velocity potential more than twice that of an adult brown trout, and almost four times that of an adult herring. (It should be noted that the original sources in the Bell figures are not known nor cited. Designers should seek studies performed for the specific species of interest. The figures are only for comparative purposes.)

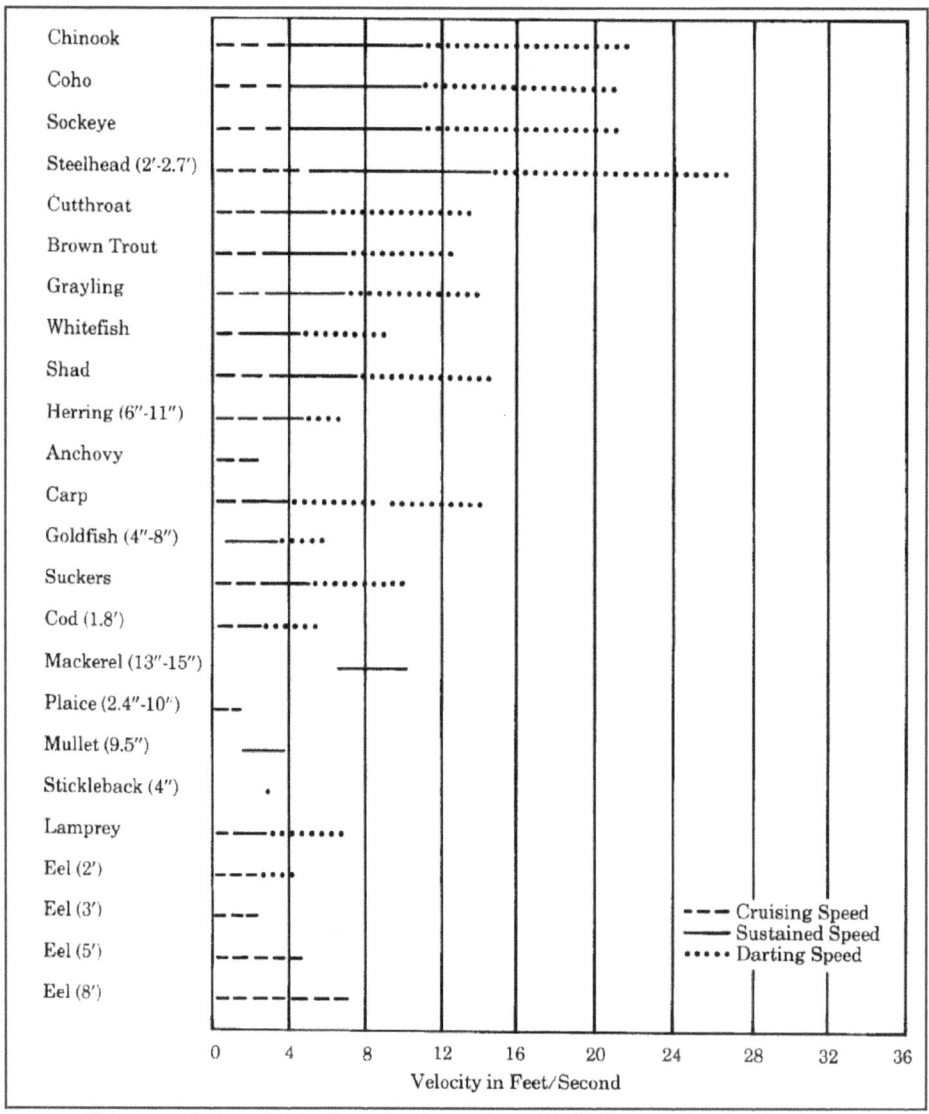

Figure 4.1. Relative Swimming Abilities of Adult Fish.

Even within a given species, there can exist a large variation between individual capabilities. This can be the result of life stage, condition or individual prowess. Figure 4.2 depicts a similar collection of swimming abilities for young fish from Bell (1986). If passage for these life stages is required, velocities thresholds drop significantly. For example, a young Coho salmon can reach sustained speeds up to 2 ft/s (0.6 m/s), while an adult is able to sustain almost 11 ft/s (3.4 m/s) . Individual fish will also exhibit dissimilar swimming capabilities, resulting in the velocity ranges depicted in Figures 4.1 and 4.2. This has serious ramifications for the selection of velocity criteria. Design for maximum swimming speed may create passage for the strongest swimmers, while maintaining a barrier to average or weak swimming individuals. Design for the weakest swimming fish will create a structure that is quite conservative.

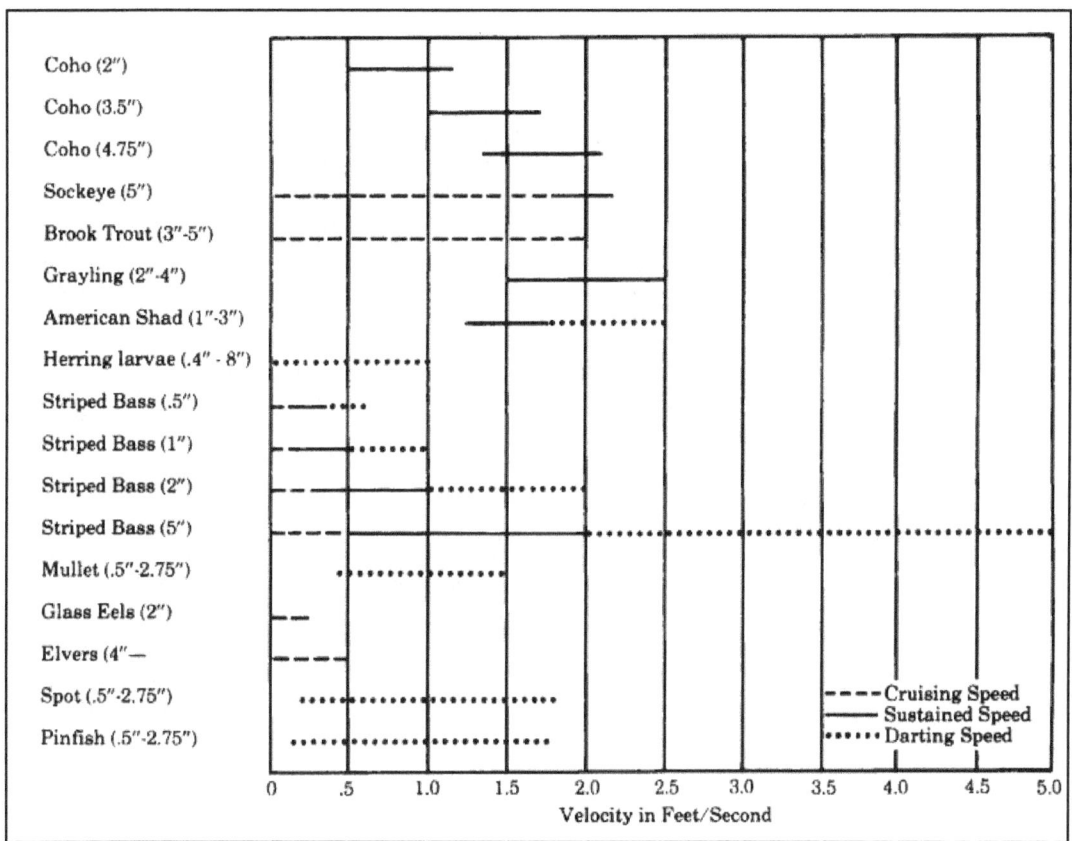

Figure 4.2.   Relative Swimming Abilities of Young Fish.

4.1.3 Depth

Fish require a minimum depth of flow to allow them to reach swimming potential (Dane, 1978). Total submergence eliminates a fish's risk of oxygen starvation, allows the fish to create maximum thrust, and lowers the risk of bodily injury through contact with the culvert bottom (Forest Practices Advisory Committee on Salmon in Watersheds, 2001). For example, Table 4.2 from Everest et al. (1985) summarizes depth requirements for a variety of salmonid and trout species from Washington and Oregon. It may be noted that fish may not be able to migrate long distances at the depths listed in the table. Data for other species and regions is under development, but not yet available.

Table 4.2. Minimum Depth Criteria for Upstream Passage of Adult Salmon/Trout.

Fish Species	Minimum Depth (ft)	Minimum Depth (m)
Pink Salmon	0.59	0.18
Chum Salmon	0.59	0.18
Coho Salmon	0.59	0.18
Sockeye Salmon	0.59	0.18
Spring Chinook	0.79	0.24
Summer Chinook	0.79	0.24
Fall Chinook	0.79	0.24
Steelhead Trout	0.79	0.24

Depth requirements vary with species and life stage, and are generally much more conservative than studies suggest. For example, Alaska requires that depth be greater than 2.5 times the depth, D, of a fish's caudal fin, as depicted in Figure 4.3 (adapted from Alaska Department of Fish and Game and Alaska Department of Transportation, 2001). The Washington Department of Fish and Wildlife specifies a minimum depth of 0.8 ft (0.24 m) for Adult Trout, Pink and Chum Salmon, and a depth of 1.0 ft (0.30 m) for adult Chinook, Coho, Sockeye or Steelhead (Bates et al. 2003). Maine employs a depth requirement of 1.5 times body depth (Maine Department of Transportation, 2004).

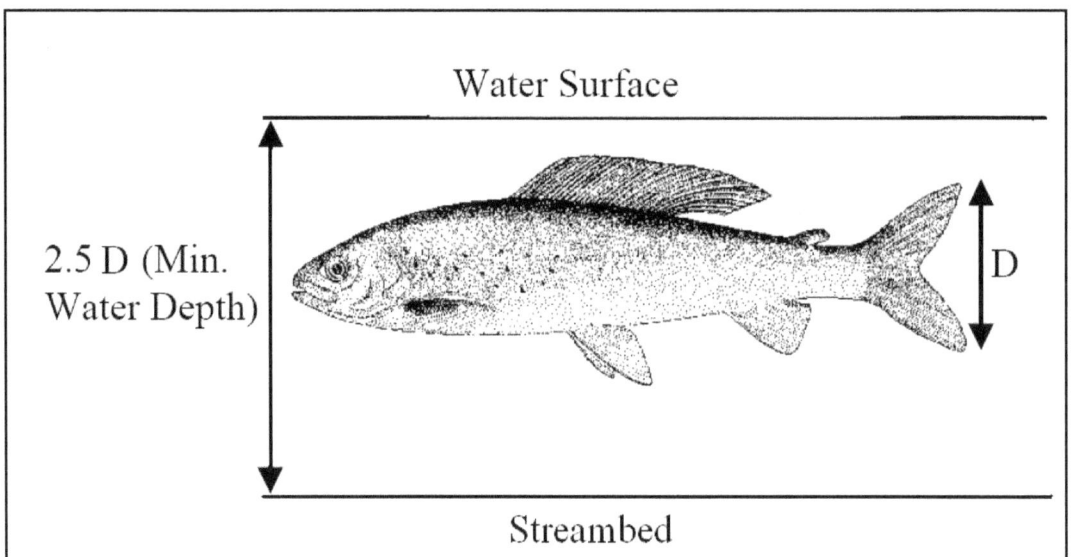

Figure 4.3. Minimum Water Depths for Fish Passage in Alaska.

4.1.4 Exhaustion

Exhaustion is a function of the rate and duration of energy expenditure. Exhaustion criteria have been experimentally derived for a variety of fish species, allowing the development of culvert velocity thresholds. Table 4.3 from Washington's fish passage manual (Bates et al., 2003) demonstrates how exhaustion and swimming speed criteria can be used to create relationships between allowable length and velocity based on fish species. In Washington State, adult trout represent a conservative lower design threshold, and are considered the species of concern in any area where specific fish species presence has not been determined.

Table 4.3.   Fish Passage Design Criteria for Culvert Installations.

Culvert Length (ft)	Culvert Length (m)	Adult Trout > 6 in (150 mm)	Adult Pink or Chum Salmon	Adult Chinook, Coho, Sockeye or Steelhead
		Maximum velocity, ft/s (m/s)		
10 – 60	3-18	4.0 (1.2)	5.0 (1.5)	6.0 (1.8)
60 – 100	18-30	4.0 (1.2)	4.0 (1.2)	5.0 (1.5)
100 – 200	30-61	3.0 (0.9)	3.0 (0.9)	4.0 (1.2)
> 200	> 61	2.0 (0.6)	2.0 (0.6)	3.0 (0.9)
		Minimum water depth, ft (m)		
		0.8 (0.24)	0.8 (0.24)	1.0 (0.30)
		Maximum hydraulic drop in fishway, ft (m)		
		0.8 (0.24)	0.8 (0.24)	1.0 (0.30)

4.2   MIGRATION AND MOVEMENT

Movement of fish populations will depend on fish species and life stage. In the Pacific Northwest, for example, adult salmon and steelhead migrate in the fall and winter months, while juvenile salmon generally out-migrate in the spring as fry and in the fall as fingerlings (Bates et al., 2003). Culvert designers in Maine must consider spawning movement of Atlantic salmon from May to November (Maine Department of Transportation, 2004). In addition, resident fish may require movement at any time of the year (Kahler and Quinn 1998; Gowan et al. 1994).

4.2.1   Anadromous Fish

Anadromous fish, such as salmon, migrate to the ocean to feed and grow, and return upstream as mature adults to spawn. Upstream movement is triggered by time of year, flow events, and a number of other environmental factors. For example, the upstream migration of spawning salmon is hypothesized to be in response to maturation, the changing length of days, and temperature regimes (Groot and Margolis 1991). Recognition of the importance of seasonal spawning runs to anadromous fish persistence led to the development of early fish passage guidance documents, e.g. Baker and Votapka (1990); Gebhards and Fisher (1972); and Evans and Johnston 1972). These migrations often occur over large distances, and the physical prowess of the individual fish degrades substantially over the course of its migration.

4.2.2   Juvenile and Resident Fish

Of more recent concern is the migration of resident and juvenile fish, e.g. Bates et al. (2003), FSSWG (2008), Robison, et al. (1999), and Admiraal and Schainost (2004). Previous knowledge held that resident populations remained fairly stationary throughout the year (Gerking, 1959); however, movement of both juvenile salmon and resident trout has been observed in response to a variety of environmental factors (Gowan et al. 1994). This includes up and down stream movement in response to extreme flows, stream temperatures, predation, lower population densities or search for food or shelter (Robison et al. 1999; Kahler and Quinn 1998; Schaefer et al. 2003).

4.2.3 Fish Presence

The distribution of fish species, life stage and migration timing is available from sources such as State and Federal Agencies, Tribal governments, commercial landowners, and non-profit organizations. Studies to ascertain fish presence may focus on larger waterways, providing low-resolution distribution maps that neglect smaller streams (Clarkin, et al., 2003).

Regional fish presence criteria may be useful, for example, fish may be assumed absent in some streams with gradients above 20 percent. To ensure that fish presence is adequately understood, some guidelines begin with the default assumption that passage is required for the weakest swimming fish contained in their criteria, e.g. Bates et al. (2003) and Robison et al. (1999). Although fish may not appear during a survey, it doesn't mean they don't inhabit the reach at some times of the year. Fish are often in areas where biologists do not expect them, and it is likely desirable to provide passage for native migratory fish that are or were historically present at the site (Clarkin et al., 2003). Assessments should be conducted when fish presence is most likely expected.

CHAPTER 5 - PASSAGE HYDROLOGY

Crossings should allow fish passage for a range of flows corresponding to the timing and extent of fish movement within the channel reach. This chapter discusses seasonality and delay, design hydrology, and flow duration curves.

5.1 SEASONALITY AND DELAY

The timing and extent of fish presence can vary from watershed to watershed (Scott and Crossman 1973), and in-stream flows may show great disparity with timing of fish migration. In addition, the presence of multiple fish species can quickly complicate evaluation of fish passage hydrology on a species by species basis. Figure 5.1 depicts the general timing of fish spawning migrations for a number of freshwater species based on biological data from Scott and Crossman (1973) (adapted from Hudy 2006). Determining species presence and sensitivity within a stream reach requires site-specific knowledge and consultation with a local fisheries biologist.

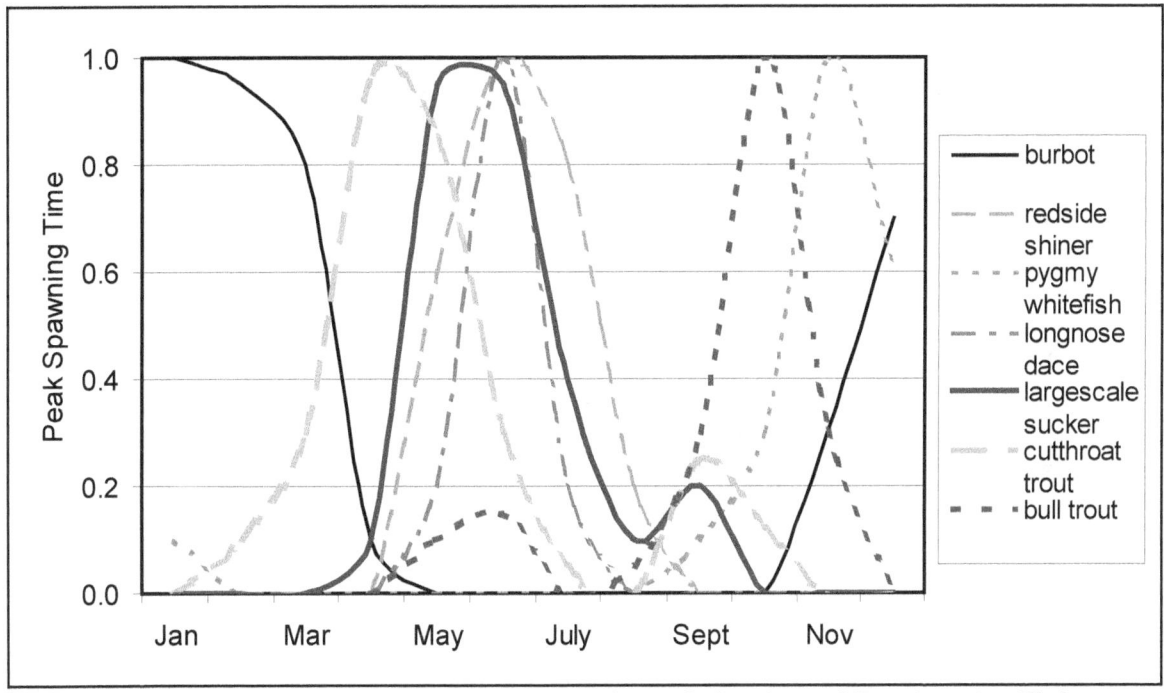

Figure 5.1. Peak Spawning Periods for a Selection of Freshwater Fish.

Certain low and high flows may prevent passage both in natural channels and culverts. Fish may be able to adapt to short interruptions to passage without negative consequences. The extent of this "allowable delay" depends on the timing and motivations for fish movement. A resident fish may be able to tolerate a short delay without extreme consequences, while a delay of a few days may be detrimental to spawning salmon, whose migrations involve significant physical changes, including a rapid depletion of fat and protein reserves (Groot and Margolis 1991). The delay caused by a single culvert can be compounded by a series of culverts that present short delays, making it imperative to understand a crossing's place in the overall watershed context. Delay has a number of negative consequences including stress and physical damages, susceptibility to disease and predation, and reduction in spawning success (Ashton 1984).

5.2 DESIGN HYDROLOGY

The design procedure described in Chapter 7 employs three design flows: flood peak, high passage flow, and low passage flow. Site-specific considerations based on seasonality and delay, as discussed above, may be used to refine selection of appropriate design flows.

5.2.1 Flood Peak, Q_p

The peak design flow is used to estimate an initial size and type of culvert based on the site-specific flood criteria. Common flood recurrence intervals range from the 25-yr to the 100-yr event and fish passage is not considered during these events. The peak design flow is determined using acceptable hydrologic methods for the appropriate recurrence interval. Acceptable hydrologic methods for determining the peak design flow include the following approaches. The designer is referred to FHWA Hydraulic Design Series 2 "Highway Hydrology" (McCuen, et al., 2002) for more information.

1. Gage data. A flood frequency analysis can be performed on gage data if:

   a. The gage is within reasonable proximity and on the same stream as the culvert.

   b. There is an adequate population of data points.

   Flood frequency analysis can be performed according to USGS Bulletin #17B: "Guidelines for Determining Flood flow Frequency". The methodology presented in Bulletin #17B is found in many software packages, such as HEC-SSP (Bruner and Fleming, 2009) and PeakFQ (Flynn, et al., 2006). Bulletin #17B uses the Log Pearson Type III distribution method. If another documented method is found to be more suited to the specific drainage then it may be used instead.

2. Regression Equations. USGS or local (State, County, etc.) regional regression equations can be used, provided the inputs can be determined. Typical inputs include average annual rainfall, terrain characteristics, average basin slope, etc.

3. NRCS Graphical Peak Discharge Method. This approach requires precipitation, computation of a time of concentration and determination of a curve number based on land use/cover and soil types.

4. Area Ratio Method. If a nearby basin with similar physical and hydrologic characteristics has gage data, the Log Pearson Type III analysis of this gage is completed. Then, an area ratio factor may be applied to determine peak flows for the design basin when the two areas are within 25 percent of each other.

   $$Q_D = Q_G (A_D/A_G)^c$$

   where Q_D is the peak flow at the design point; Q_G is the peak flow at the gage; A_D is the drainage area at the design point; A_G is the drainage area at the gage; and c is drainage area exponent from the applicable regression equation.

5. Recently published and verifiably accurate peak flows (i.e. FEMA FIS or other). If peak flows are available, but the Q_P for the culvert is not available, interpolation or extrapolation on log-probability paper is an acceptable method for determining Q_P. For example, if the Q_P is the 50-year design flow, and the 10-year, 25-year, and 100-year return interval floods are known, then log-probability interpolation or extrapolation on a frequency plot is an acceptable method for determining Q_P (See Figure 5.2).

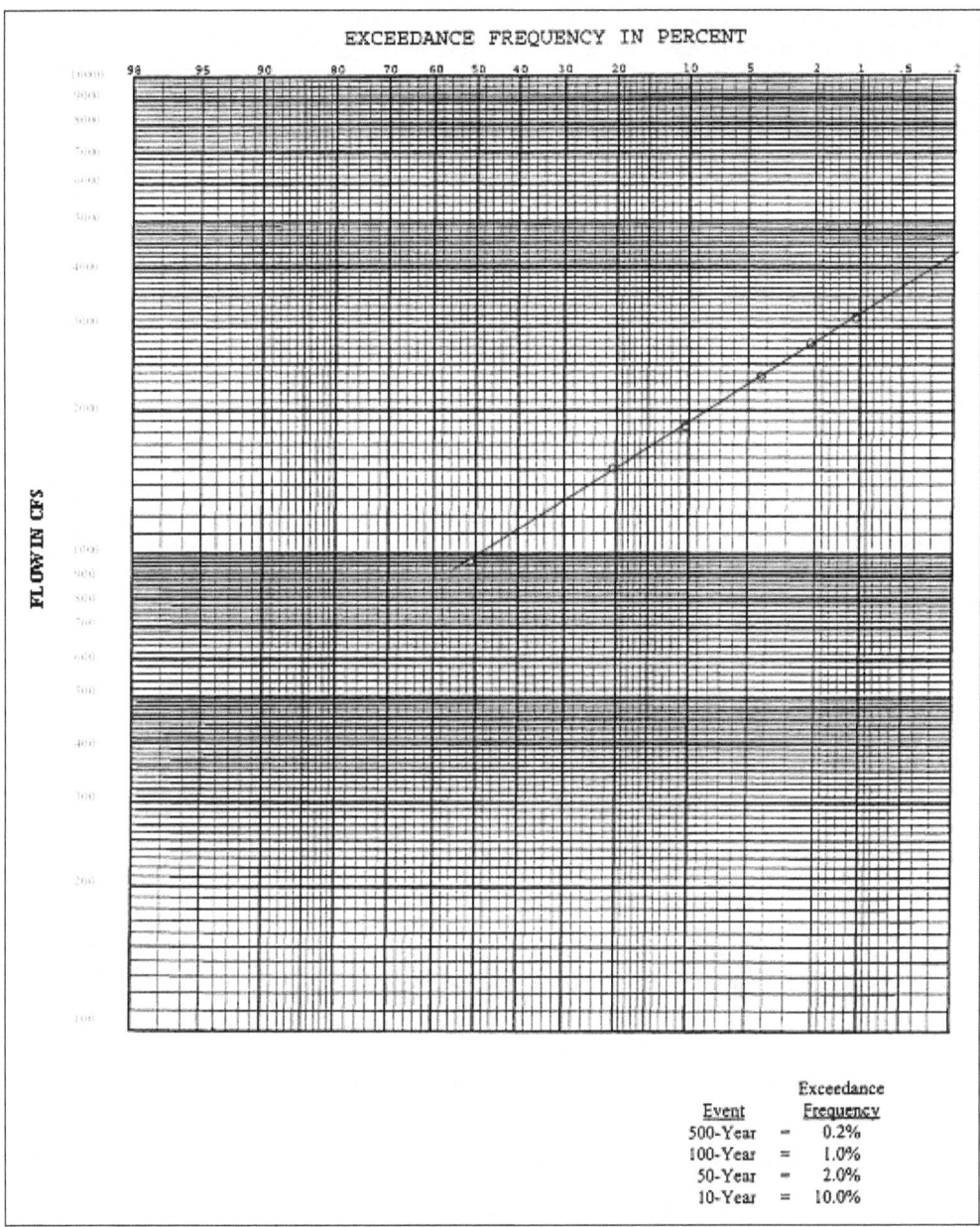

Figure 5.2. Example Log-Probability Plot.

6. <u>Unit Hydrograph</u>. If none of the previous methods are viable, unit hydrograph techniques can be used to synthesize various return-interval floods using precipitation data and basin characteristics. TR-20 and HEC-HMS are examples of software tools that can perform this task.

5.2.2 High and Low Passage Flows

In a natural stream reach, fish respond to high flow events by seeking out shelter until passable conditions resume (Robison et al. 1999). During extreme low flows, shallow depths may cause the channel itself to become impassable (Clarkin et al. 2003; Lang et al. 2004). Generally, upper and lower thresholds bound the flow conditions at which fish passage must be provided and these are defined here as the high and low passage flows.

High passage flow, Q_H, represents the upper bound of discharge at which fish are believed to be moving within the stream, while low passage flow, Q_L, is the lowest discharge for which fish passage is required, generally based on minimum flow depths required for fish passage. High passage and low passage design flows are not defined in the same manner throughout the country. This variation may reflect differences in hydrology and fish species from region to region, but it may also reflect inconsistencies in defining these terms. Tables 5.1 and 5.2 summarize alternative definitions for high and low passage flows (adapted from Hotchkiss and Frei, 2007).

Table 5.1. State and Agency Guidelines for Q_H.

State/Agency*	Guideline
Alaska	Q_2D_2: the average discharge 24 hours before and after the 2-yr flood. May be estimated as 40% of the Q_2. (Guideline was developed for Southeast Alaska.)
California Dept. of Fish and Game	Standards vary from 1 to 10% annual exceedance flow for various groups of fish.
California Dept. of Transportation (2007)	Varies from 1 to 10% annual exceedance flow, species/life stage specific. May be estimated as 50% to 10% of the Q_2.
Idaho	Less than 2-day delay during period of migration.
NMFS NW Region	5% exceedance flow during period of upstream migration.
NMFS SW Region	For adult salmon and steelhead 1% annual exceedance flow or 50% Q_2. For juveniles, 10% annual exceedance flow.
Oregon	10% exceedance flow during migration period, species specific. Approximate by $Q_{10\%} = 0.18*(Q_2) +36$ ft^3/s (1.0 m^3/s) where $Q_2 > 44$ ft^3/s (1.2 m^3/s). Where $Q_2 < 44$ ft^3/s (1.2 m^3/s), use Q_2.
Vermont (Bates and Kirn, 2007)	Regression equations for a flow with 20% exceedance probability for 2 consecutive days in April for Spring flow and 2 consecutive days in November for Fall flow.
Washington	10% exceedance flow during migration period, species specific.

*All guidelines from Clarkin, et al. (2003) unless otherwise noted.

Table 5.2. State and Agency Guidelines for Q_L.

State/Agency*	Guideline
California Dept of Fish and Game	Standards vary from 50-95% annual exceedance flow for various groups of fish.
California Dept. of Transportation (2007)	Varies from 50 to 95% annual exceedance flow, species/life stage specific. Alternative minimum ranges from 1 to 3 ft^3/s (0.028 to 0.085 m^3/s).
NMFS NW Region	95% exceedance flow during months of upstream migration
NMFS SW Region	Adult Salmon: greater of 3 ft^3/s (0.085 m^3/s) or 50% exceedance flow. Juveniles: greater of 1 ft^3/s (0.028 m^3/s) or 95% annual exceedance flow
Oregon	2-yr, 7-day low flow (7Q2) or 95% exceedance flow for migration period, species specific.
Vermont (Bates and Kirn, 2007)	2-yr, 7-day low flow (7Q2). May be estimated as 0.139 ft^3/s/mi^2 (0.00152 m^3/s/km^2) times the drainage area.
Washington	2-yr, 7-day low flow (7Q2). Natural bed culverts must be maintained to ensure low-flow channels are ok.

*All guidelines from Clarkin, et al. (2003) unless otherwise noted.

Although there appears to be a broad array of approaches for estimating Q_H and Q_L, the result is two numbers representing a range of flows between which fish may be expected to move. By considering this range of flows in AOP design, passage is evaluated for low and high flows. The desired range of flows may expand or contract somewhat depending on the species, life stage, or season of particular interest, but the ultimate objective is to derive the appropriate range of passage flows for culvert design.

As indicated in Table 5.1, Q_H has historically been defined as a specific exceedance probability quantile based on either an annual or seasonal flow duration curve. (See section 5.3 for a description of flow duration curves.) In some cases, such as the NMFS SW Region, these quantiles can be estimated as a percentage of the Q_2 flood level.

As indicated in Table 5.2, Q_L has also historically been defined as a specific exceedance probability quantile based on either an annual or seasonal flow duration curve or as the 7Q2 statistic. In some cases, minimum flows of 1 to 3 ft^3/s (0.028 to 0.085 m^3/s) are specified.

Available methods for estimating Q_H and Q_L follow. The most appropriate method or methods depends on the site-specific situation and the availability of supporting data.

1. <u>Developing exceedance probability quantiles and/or 7Q2 from daily gage flow data at or near the culvert</u>. At least ten years of data are necessary to support reasonable estimates. Consideration should be given to whether or not the period of record corresponds to a particularly dry or wet period. A flow duration curve is created to estimate the desired exceedance probability quantiles. If the 7Q2 is needed a rolling 7-day averages are analyzed statistically to generate the 7Q2.

2. <u>Developing exceedance probability quantiles and/or 7Q2 from daily gage flow data from gage data in similar watersheds</u>. This approach is for the more common situation of an ungaged culvert site. A minimum of ten years of data is necessary and daily flows are adjusted based on the drainage areas of the gage and the site. Multiple similar watersheds could be used for this purpose.

3. <u>Regression equations</u>. Regression equations may be useful for ungaged watersheds when they have been developed for the region and characteristics of the culvert location. The practice of Vermont, for example, is to calculate Q_L as a linear relationship to drainage area may be considered as a regression equation (Bates and Kirn, 2007). Washington State, on the other hand, has separate regression equations for Q_H for watersheds West and East of the Cascades (Powers and Saunders, 1996, and Rowland et al. 2003).

4. <u>Fraction of a peak flood statistic, e.g. Q_2</u>. Q_H is estimated in some locales as a fraction of the Q_2 flood statistic. There is little reason to suspect that lower flow statistics can be estimated from a high flow statistic. However, if the first three approaches are unavailable, this may be the only alternative to quantify fish passage flows.

5.3 FLOW DURATION CURVES

Flow duration curves (FDCs) are one method for estimating Q_H and Q_L and are commonly used to graphically illustrate streamflow characteristics for a gaged location. Shown in Figure 5.3 (Lang, et al., 2004), an FDC displays the percent of time the indicated discharge is exceeded. FDCs may be created to represent the entire period of record, or for a portion of a water year (or season).

FDCs are most often developed for the entire period of record. If done for a particular water year (e.g. wet or dry), the FDC is called an annual flow duration curve. Annual flow duration curves are being used more frequently because a return period or chance of occurrence may be assigned to any observed water year within a period of record (Castellarin, et al., 2007).

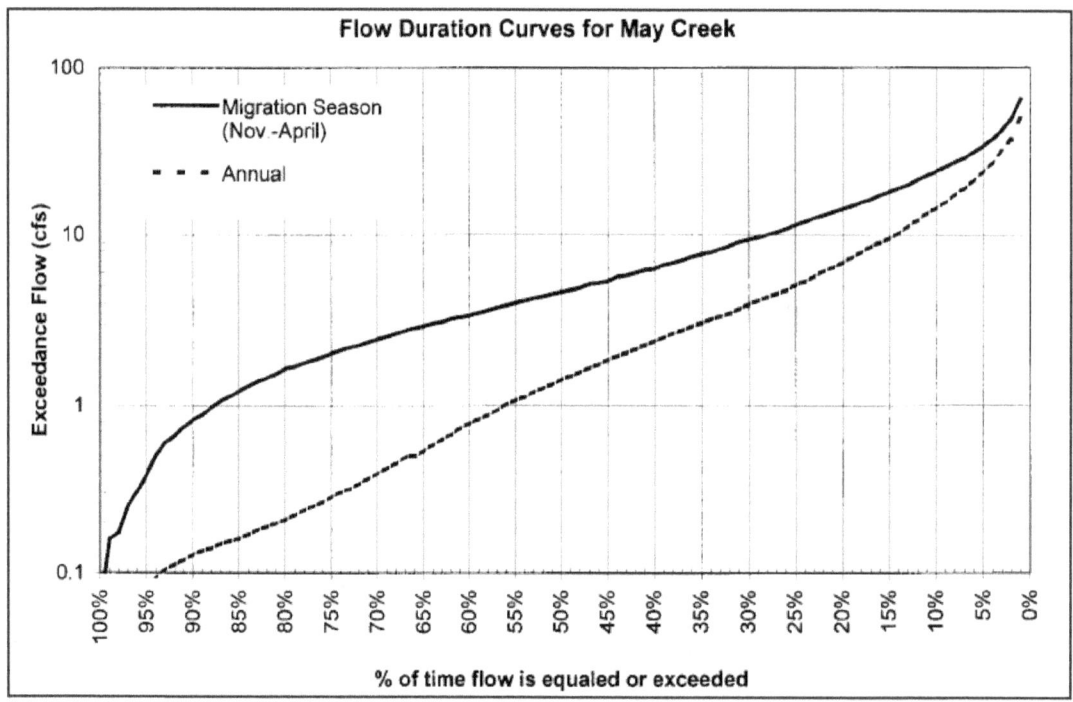

Figure 5.3.   Flow Duration Curve for an Annual and a Seasonal Time Period.

There have been many efforts to develop FDCs for ungaged stream locations. All methods use stream gage information as a basis for extrapolating to ungaged catchments. Most methods relate key points of the FDC to watershed characteristics, allowing a user to define an FDC for an ungaged area by calculating the relevant watershed characteristics and then finding the ungaged FDC using regression equations. The equations are either related to the observed FDCs or are the result of simulating observed FDCs with fitted probability distributions (Archfield et al., 2007, Fennessey and Vogel, 1990).   One procedure (Studley, 2000) combines miscellaneous discharge measurements at an ungaged site to find an FDC for the location by 'scaling' the measurement to a nearby stream gage. Another method (Archfield and Vogel, 2008) develops FDCs for ungaged sites by computing key points on the curve using regression equations and filling in other FDC points using equations based on the key points obtained from regression. Yet another method (Doyle et al., 2007) uses XPSWMM to generate a continuous synthetic streamflow record for a site and derives an FDC from the simulated record.

The Massachusetts District of the U.S. Geological Survey is developing a method of determining FDCs based on the period of record from 1960 – 2004 (Archfield, 2007).   This method will be added to the StreamStats program for Massachusetts. It is estimated that it will take several years to develop similar procedures for all States.

CHAPTER 6 - STREAM GEOMORPHOLOGY

As a rigid structure in a dynamic environment, culverts must be designed with channel processes in mind. Effective designs consider the channel and watershed context of the crossing location. Channels are continually evolving, and an understanding of stream adjustment potential must be addressed. Without proper consideration, well-intended plans could detrimentally affect the stream system and related habitat (Castro, 2003; Furniss, 2006).

6.1   CHANNEL CHARACTERISTICS

6.1.1 Channel Width

Channel width is defined in a variety of ways depending on the application. Bankfull width and active channel width are commonly cited width parameters. Figure 6.1, from Taylor and Love (2003), schematically represents these two concepts.

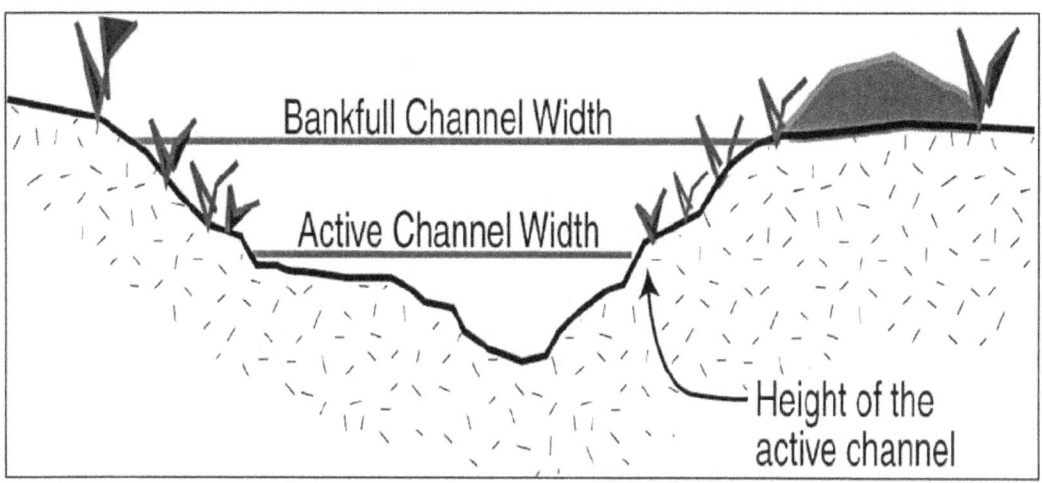

Figure 6.1.   Bankfull and Active Channel Widths.

6.1.1.1 Active Channel Width

The active channel may be identified by the ordinary high water (OHW) mark, that is, the elevation delineating the highest water level that has been maintained for a sufficient period of time to leave evidence on the landscape (Taylor and Love 2003). Other representations may include erosion, shelving or terracing, change in soil characteristics, a break or destruction of terrestrial vegetation, moss growth on rocks along stream margins, vegetation changes from predominantly aquatic to predominantly terrestrial, or the presence of organic litter or debris (Taylor and Love, 2003; Bates et al., 2003).

6.1.1.2 Bankfull Width

Bankfull width is the water top width at the bankfull discharge. Bankfull discharge is that flow rate that fills a channel to the point of overflowing onto the floodplain. Generally this definition presumes the channel is in equilibrium and not incising because bankfull width is considered to be a natural equilibrium dimension for the channel. Other descriptions of discharges that strongly influence channel characteristics, including width, are channel-forming discharge, dominant discharge, and effective discharge (see Glossary). All of these terms attempt to describe a discharge that is directly associated with equilibrium channel conditions. However, the utility of each definition is limited because it is not only a range of flows that strongly influence channel characteristics, but also, the bed and bank materials, watershed uses, vegetation, and other factors.

Bankfull discharge is often characterized as a 1- to 2-year event, when flow within the channel just begins to spill over into the active floodplain (Leopold et al., 1964), though the appropriate return period can be 5- or 10-yrs in some semi-arid and arid environments. When floodplains are absent or difficult to ascertain, as in entrenched mountain streams, markers used to determine bankfull and active channel show little variation (Bates et al. 2003). Difficulty in determining bankfull flow in the field prompts some to provide guidelines for estimation of bankfull width based on surveyed cross sections and return period flow, e.g. Maine Department of Transportation (2004). This type of estimation may show great disparity when compared with field observations of channel-bed width (Mussetter, 1989).

6.1.2 Gradient

Channel degradation can require channel modification, or considerations of the impact of increased slope on channel stability, substrate, and future conditions (Robison et al. 1999; FSSWG, 2008; Bates et al., 2003). AOP through a culvert is more likely to be successful when culvert bed slopes are consistent with the slopes of the adjacent stream channel. Oversized sediment may be utilized to provide more leeway with regards to stream slope.

6.1.3 Bed Material and Embedded Culverts

The benefits of natural streambeds and embedded culverts are widely recognized in AOP applications, e.g. Venner Consulting and Parsons Brinkerhoff, 2004; Bates et al., 2003; Taylor and Love, 2003; and Clarkin et al., 2003). Bed material provides barrel roughness, which provides areas of low velocity that may be conducive to passage, mimics natural hydraulics, and is self-sustaining when designed properly (White, 1997).

6.1.4 Key Roughness Elements

Many designers incorporate key roughness elements in passage designs, e.g. Robison et al., 1999; FSSWG, 2008; and Browning, 1990. Such features are intended to increase bed stability and provide resting areas and hydraulic diversity to a crossing conducive to passage. Key roughness elements may use any number of materials including oversized substrate, constructed channel features including banks, stone sills, boulder clusters, log sills, and baffles.

6.2 CHANNEL TRANSFORMATIONS

6.2.1 Channel Evolution

Most stream channels are constantly changing making it necessary to assess whether a particular channel cross-section and slope observed at a particular time is characteristic of dynamic equilibrium for the channel or if the channel is evolving to a new state of dynamic equilibrium. Channels that are in dynamic equilibrium may be set on an evolutionary path by large hydrologic events or human changes to the channel and watershed.

For example, Figure 6.2 (Schumm, et al., 1984) depicts channel evolution from a stable state (dynamic equilibrium up to the 2-yr discharge event) through several unstable states to a new stable state. In cross-section I, the channel is stable because the bank height is less than a critical bank height for the stream. The evolution is set in motion by a 10-yr storm that causes the stream to incise resulting in an increase in the bank height to an unstable dimension. In cross-section III, the bank is failing. The stream widens to reduce the flow depth and shear stresses on the streambed. A new channel forms (cross-section V) under altered conditions of dynamic equilibrium.

Channel evolution, as shown in Figure 6.2, creates challenges for determining bankfull dimensions of a given stream. During the evolutionary period, field measurements of bankfull width would not be appropriate because the channel is not stable and the dimensions would not be representative.

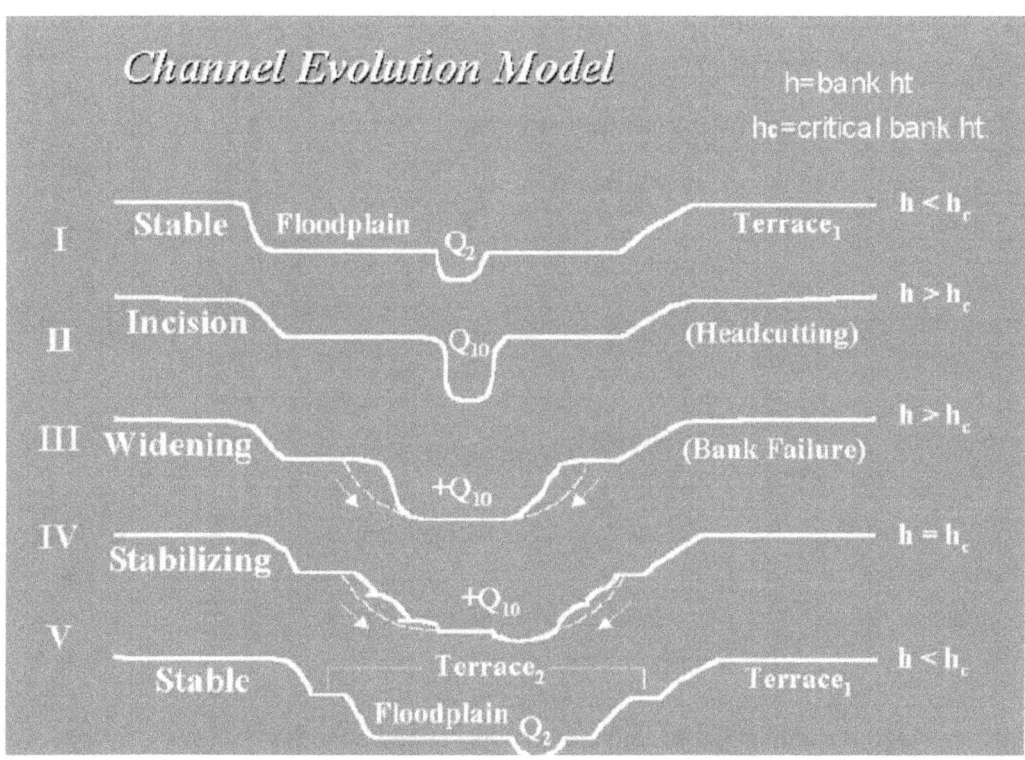

Figure 6.2.   Channel Evolution Model.

6.2.2 Channel Incision, Headcuts, and Aggradation

As channels continually evolve and migrate, channel adjustment can lead to structure failure. Installations that fail to recognize channel processes may compromise fish passage and alter the quantity and quality of stream corridor habitat (Castro, 2003).

In situations where a current culvert installation is acting as a control point, removal and replacement with a larger structure (or lowering the invert) may allow channel incision to progress upstream uncontrollably, or until another control point is reached. Stream reaches actively aggrading or incising may cause culverts to be ineffective for passage.

6.3 STREAM CLASSIFICATION

Systems for stream classification are useful tools in building awareness of stream form and function. Methods may describe the channel in terms of cross-sectional shapes, morphological parts of the stream, and interactions between flow and sedimentation (Bunte and Abt 2001). The following sections introduce two stream classification methods. For more information it will be useful to examine the documents referenced below and Hydraulic Engineering Circular No. 20 (Lagasse et al. 2001).

6.3.1 Montgomery and Buffington

Montgomery and Buffington (1993 and 1998) created a stream classification system based on steeply sloped channel systems in the Pacific Northwest that is applicable to similar regions elsewhere. Their methodology follows changes in channel morphology as steep forested headwater streams run through steep valleys and hill slopes, gentle valleys, and eventually low gradient valleys (Bunte and Abt 2001). As water flows to the ocean, channel types generally transition from cascade to step-pool, plane bed, pool-riffle, and dune-ripple. Channel bedform is described by the type and size of sediment, sediment transport capabilities, and hydraulic conditions within a stream reach. Taken from Bunte and Abt (2001), Table 6.1 summarizes this classification system with respect to channel geomorphic and hydraulic conditions.

Table 6.1.   Stream Classification by Montgomery and Buffington.

Stream Gradient, ft/ft (m/m)	Stream Type	Typical Bed Material	Dominant Sediment Source	Dominant Sediment Storage	Typical pool spacing*
0.03 – 0.20	Cascades	Cobble-boulder	Fluvial, hill slopes, debris flows	Around flow obstructions	< 1
0.02 – 0.09	Step-pool	Cobble-boulder	Fluvial, hill slopes, debris flows	Bedforms	1 - 4
<0.02 – 0.05	Plane-bed, forced pools	Gravel-cobble	Fluvial, bank failure, debris flows	Overbank	None
<0.001 – 0.03	Pool-riffle	Gravel	Fluvial, bank failure	Overbank, bedforms	5 – 7
< 0.001	Dune-ripple	Sand	Fluvial, bank failure	Overbank, bedforms	5 – 7

*multiple of channel width.

A reach-scale categorization allows streams to be categorized based on relative positions within the watershed and sediment transport characteristics. This type of analysis is useful in understanding the potential response of a channel reach to a crossing installation. Montgomery and Buffington define reach level morphologies as source, transport and response reaches (Montgomery and Buffington 1993).

Source reaches contain as much or more sediment than the stream can transport. Transport reaches are high gradient supply-limited channels, which are unlikely to respond quickly or severely to disturbance. This includes bedrock, cascade and step-pool channels. Response reaches are lower gradient transport-limited channels with a high potential for morphological adjustment in response to sediment input. This general classification covers plane-bed, pool-riffle and braided channels. The transition from transport to response reach is where the impacts of increased sediment supply will have the largest impact, as sediment supplied by the transport reach will readily settle out at the first reach that cannot maintain sediment transport capacity (Montgomery and Buffington 1993).

A crossing location within a particular reach, as well as the proximity of other reaches will help a designer ascertain the potential geomorphic response of the stream. Crossings that fall at the intersection of two different channel types, for example, could indicate channel incision, or that the crossing is located at a point of geomorphic transition (FSSWG, 2008). Crossings placed in a response reach typically will require consideration of channel processes and morphological impacts, including channel aggradation and lateral movement.

6.3.2 Rosgen

Rosgen channel classification is based on five morphometric parameters of the channel and its floodplain: entrenchment ratio, width-depth ratio at bankfull flow, sinuosity, stream gradient, and mean bed particle size (Rosgen, 1994; Rosgen, 1996). (Entrenchment ratio is the ratio of the floodprone width to the bankfull width of the channel. The floodprone width is measured at an elevation such that the floodprone depth is twice the bankfull depth.)

These characteristics are used to distinguish seven stream types, represented by capital letters A to G. Taken from Bunte and Abt (2001), Table 6.2 lists the morphological characteristics of Rosgen's stream types.

Table 6.2.   Morphological Characteristics of the Major Rosgen Stream Types.

Stream Type	Morphological Characteristics
A	**Step-pool or cascading**: plunge and scour pools, high energy, low sediment storage, stable.
B	**Riffles and rapids**: some scour pools, bars rare, stable.
C	**Pool-riffle sequences**: meandering, point bars, well-developed floodplain, banks stable or unstable.
D	**Braided**: multiple-channels, shifting bars, scour, deposition, high sediment supply, eroding banks.
DA	**Anastomosing**: multiple channels, pool-riffle, vegetated floodplain, adjacent, wetlands, stable banks.
E	**Meadow meanders**: well-developed floodplain, riffle-pool, relative high sediment conveyance.
F	**Valley meanders**: incised into valleys, poor floodplain, pool-riffle, banks stable or unstable.
G	**Gullies**: incised into hill slopes and meadows, high sediment supply, unstable banks, step-pool.

Channels can be further distinguished using numbers to represent bed material and particle size, and lowercase letters to represent deviation from expected channel slopes. For example, a stream classified as C4b is a C-type stream with a gravel bed and gradient within the range of 0.02-0.039, which is more typical of a B-type stream (Rosgen 1994). Accurate classification requires a longitudinal and cross-sectional channel survey and sediment sample analysis.

6.3.3 Summary

Stream classification systems may be useful in understanding basic channel reach geometry and dominant geomorphic processes, which can be valuable in predicting channel response to modification or culvert replacement. Certain channel types can carry specific design challenges. For example, risk of floodplain constriction and/or lateral adjustment is associated with Rosgen C, D and E channels (FSSWG, 2008). As mentioned above, plane bed, pool-riffle, and dune-ripple channels are associated with response reaches, and are likely to show the most dramatic response to disturbance (Montgomery and Buffington, 1993). It is important to note that these classification systems are not always tested outside the regions and typical stream types for which they were created. For example, low gradient, highly mobile sand bed streams may require special consideration.

CHAPTER 7 - DESIGN PROCEDURE

The variables required to implement the design procedure include the following:

1. Peak design flow, Q_P. This flow may be the Q_{25}, Q_{50}, or Q_{100} required for the site to address design flood flows.

2. High passage design flow, Q_H. This is the maximum discharge used for passage design. It may apply to the entire year or to a specific season.

3. Low passage design flow, Q_L. This is the minimum discharge used for passage design. It may also apply to the entire year or to a specific season.

4. Bed material gradation. Representative bed sizes including D_{16}, D_{50}, D_{84}, and D_{95} are required. Presence or absence of an armor layer should be noted.

5. Permissible shear stress, τ_p, of the bed material.

Five fundamental tests are applied as part of the procedure. If any test is failed, design adjustments are specified. The tests are:

1. Does the culvert satisfy the peak flow requirements?

2. Is the bed material in the culvert stable (no movement or sediment inflow equals outflow) for the high passage design flow?

3. Is the bed material in the culvert stable for the peak design flow? (An anchoring layer/device below the bed material may be required to satisfy this test.)

4. Is velocity in the culvert for the high passage design flow consistent with upstream and downstream channel velocities?

5. Is depth in the culvert for the low passage design flow consistent with upstream and downstream channel depths?

Figure 7.1 provides a flow chart of the 13-step design procedure. Step 1 involves determination of the hydrologic requirements for the site for both flood flows and passage flows. The passage flows do not require determination of target species and life stages, though if they are known for a site should be used in defining the passage flows. Step 2 defines the project reach and establishes the representative channel characteristics appropriate for the design.

Because it is inadvisable to place a fixed structure, such as a culvert, on an unstable stream, Steps 3 and 4 are to identify whether the stream is stable (Step 3). If not, channel instabilities are analyzed and potentially mitigated (Step 4).

In Step 5, an initial culvert size, alignment, and material are selected based on the flood peak flow. Subsequently, the stability of the bed material is analyzed under the high passage flow (Steps 6 and 7) and flood peak flow (Steps 8 and 9). If any of the criteria are not satisfied, the designer returns to Step 5 to find an alternative culvert configuration, usually larger.

Steps 10, 11, and 12 focus on the velocity and depth in the culvert. However, these parameters are not compared with species-specific values, but rather are compared with the values upstream and downstream of the culvert insuring that if an organism cause pass the upstream and downstream channel, it will also be able to pass through the culvert. If species-specific values are relevant and available for the site, they may also be incorporated into the design.

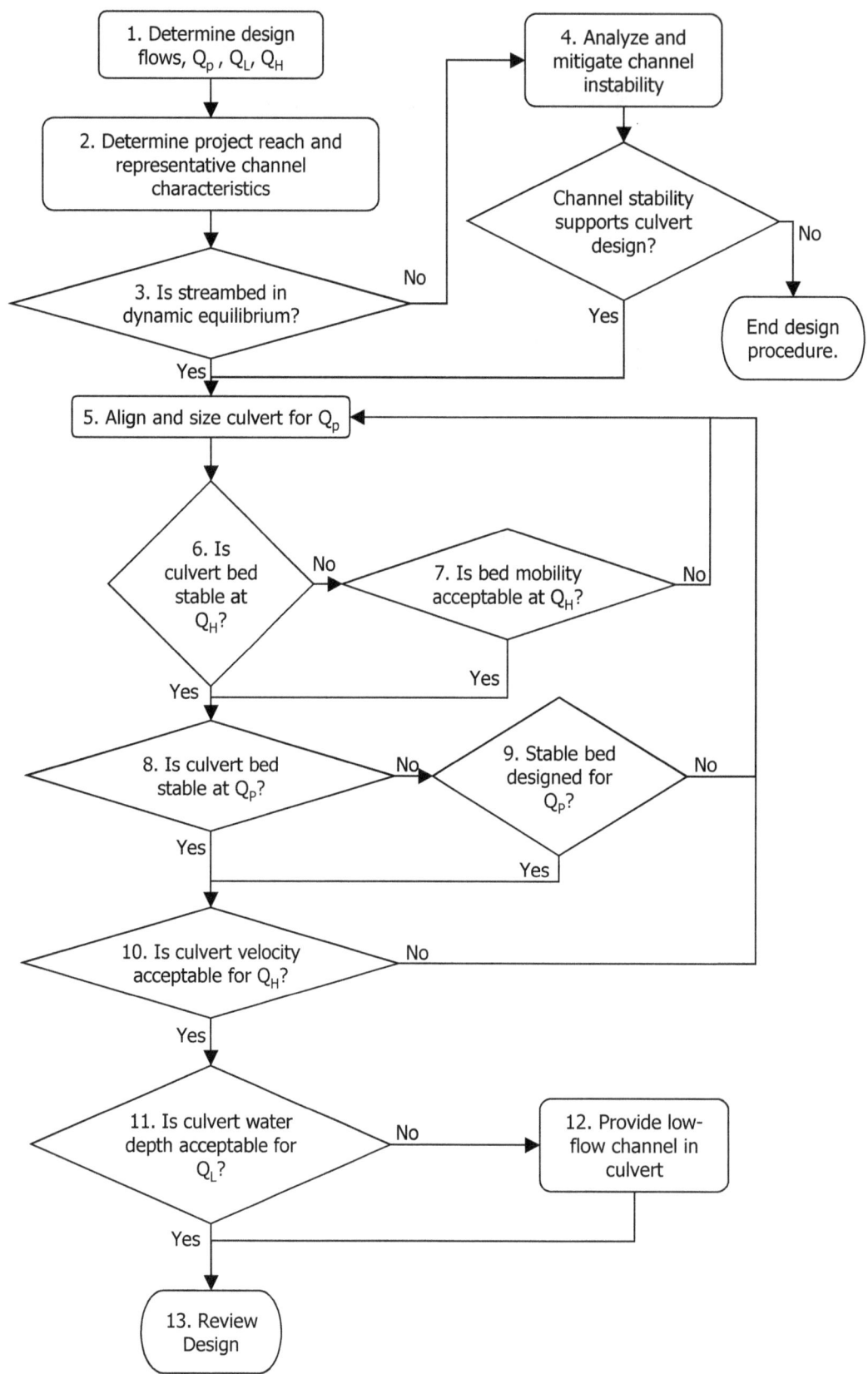

Figure 7.1. Design Procedure Overview.

Step 13 allows the designer to review the completed design for its compatibility with the project objectives, environmental requirements, and construction and maintenance costs. Since there are many culvert types, sizes, and materials available the process may include several iterations prior to selecting a final design.

7.1 STEP 1. DETERMINE DESIGN FLOWS.

The design procedure incorporates three design flows: 1) peak flow, Q_P, 2) high passage flow, Q_H, and 3) low passage flow, Q_L. As with all hydrologic analyses, hydrologic uncertainty should be considered in developing estimates. This may include assessment of climatological trends that may suggest that more recent data should be weighed more heavily than older data, for example.

The minimum recurrence interval for peak flow design is usually specified in local, state, or Federal design guidance. The appropriate recurrence interval should consider the lifecycle costs, risks and costs of failure, AOP, and other design objectives. Typically, the Q_{25}, Q_{50}, or Q_{100} peak design flow is used to size the culvert for peak flow conditions. The overtopping discharge may also be used as Q_p, especially in retrofit situations where the roadway profile cannot be adjusted, or when the peak design discharge is accommodated by allowing a portion of the flow to overtop the roadway.

The peak design flow, Q_p, is determined using acceptable hydrologic methods for the appropriate recurrence interval. The designer is referred to Chapter 5 of this document and FHWA Hydraulic Design Series 2 "Highway Hydrology" (McCuen, et al., 2002) for more information.

If Q_H and Q_L guidelines are specified for a project site those guidelines should be used and an appropriate methodology applied to quantify Q_H and Q_L. Methods for estimating Q_H and Q_L are listed in Chapter 5. (Tables 5.1 and 5.2 provide a selection of example guidelines.) Where passage design is desired, but Q_H and Q_L are not defined, or no site-specific guidance is available, the following default guidance may be used.

In the absence of site-specific guidelines, the Q_H should be defined as the 10 percent exceedance quantile on the annual flow duration curve. If development of a flow duration curve is not possible for the site and an appropriate regression equation is unavailable, Q_H should be estimated as 25 percent of the Q_2. These recommendations are based on extension of current practice as represented in Table 5.1.

Similarly, in the absence of site-specific guidance, Q_L should be defined as either the 90 percent exceedance quantile on the annual flow duration curve or the 7-day, 2-year low flow (7Q2). If both estimates are available, the smaller should be selected. Regardless of the source of the estimate, Q_L should not be lower than 1 ft³/s (0.028 m³/s). These recommendations are based on specifying an appropriate extension of current practice as represented in Table 5.2.

7.2 STEP 2. DETERMINE PROJECT REACH AND REPRESENTATIVE CHANNEL CHARACTERISTICS.

The project reach is the portion of the waterway that is the primary geographic scope of the design process. It includes the road crossing location and extends upstream and downstream to points beyond the geomorphic influence of the road crossing. The extents of the project reach are often defined by grade control features.

This step is to determine the extent of the project reach and identify representative characteristics of the upstream and downstream channel within the project reach. The characteristics of interest are channel geometry, slope, and bed material gradation.

The project reach must extend sufficiently far upstream and downstream of the culvert location to adequately identify representative characteristics for the channel, including geomorphic form and key features such as riffle-pool or chute-pool areas in order to produce meaningful comparisons with design conditions at the culvert location. At a minimum, the project reach should extend both upstream and downstream from the culvert location (referenced to the toe of the roadway embankment) no less than three culvert lengths or 200 feet (61 m), whichever is greater. Since culvert length will not be established at this point an estimate is required. The upstream and downstream extents of the project reach should be established at permanent stream grade control points, if present.

Representative channel characteristics are extracted from cross-sections within the project reach. At a minimum, six (6) cross-sections should be established: three (3) upstream and three (3) downstream of the culvert location. Cross-sections should be located considering geomorphological features. For example, it may be appropriate to locate four sections upstream to capture two pool-riffle pairs with an additional four sections downstream. Cross-sections near the inlet and outlet of the culvert, but outside of the influence of the roadway embankment, are desirable.

Bed material samples are taken in the project reach and analyzed to produce a particle size distribution curve including estimates of the D_{16}, D_{50}, D_{84}, and D_{95} of the bed. The presence or absence of an armor layer should be noted. If the channel is armored, pebble counts and bulk sampling are both vulnerable to under-representation of smaller fractions and over-representation of larger fractions. See Bunte and Abt (2001) for detailed sampling guidelines.

If the bed material is observed to be consistent throughout the project reach, one sample upstream and one sample downstream of the culvert location is sufficient. However, if variability is observed, additional samples sufficient to characterize the variation, up to one per cross-section, should be collected. If there appears to be a significant change in material type or gradation upstream of the project reach, an additional sample should be taken beyond the limits of the project reach for evaluation of sediment supply. Emphasis will be placed on gradation samples taken from locations with a slope close to the slope eventually selected for the culvert bed.

Channel slope is defined by a longitudinal profile. For fairly uniform channels with little variation between cross-sections, channel thalweg elevation at each cross-section may be sufficient to define the longitudinal profile. However, most channels will require additional thalweg elevations between the cross-sections to capture the vertical variation, including control points.

For replacement crossings, the existing crossing may have altered the "natural" characteristics at the adjacent downstream and upstream reaches. For example, the channel may be aggrading upstream or a scour hole may be present downstream, or both. (When replacing the culvert, the disposition of the deposited sediment should be considered. If the quantity is small, natural processes may be allowed to remove it. However, fishery resources and water quality should be considered for larger sediment deposits.) Cross-section locations should be established beyond the zone of influence of these localized perturbations. If the existing crossing has created control points that will be maintained in the replacement crossing, such features need to be included in the survey.

7.3 STEP 3. CHECK FOR DYNAMIC EQUILIBRIUM.

Successful application of the design procedure requires a stable channel or one that is in dynamic equilibrium. If a culvert is located in an unstable reach, HEC 20 (Lagasse, et al., 2001a) or other appropriate reference should be consulted for stream stability assessment prior to design of the crossing. If a stable reach is not present or mitigation of the instabilities is not possible or beyond the resources available for the site, then a culvert crossing may be less appropriate than avoiding the crossing or providing a bridge.

HEC 20 (Lagasse, et al., 2001a) observes that "a state of 'dynamic' equilibrium may exist if there is a long-term balance (on an engineering time scale) between the water and sediment supplied by the watershed and that transported by the stream system. When dynamic equilibrium exists, bed scour and fill and bankline migration may occur, but on an engineering time scale, reach averaged characteristics and the balance between sediment inflow and sediment outflow are maintained." Common indicators of channel instability include (Lagasse, et al., 2001a):

- Headcut: Channel degradation associated with abrupt changes in the bed elevation (headcut) that generally migrates in an upstream direction. Headcuts downstream of a culvert location will move up toward the culvert and potentially threaten the installation.

- Bank instability/erosion: An unstable bank may be indicated by steep slopes (greater than 30 percent) and a lack of woody vegetation. Active bank erosion can be recognized by falling or fallen vegetation along the bank line, cracks along the bank surface, slump blocks, deflected flow patterns adjacent to the bankline, live vegetation in the flow, increased turbidity, fresh vertical faces, newly formed bars immediately downstream of the eroding area, and, in some locations, a deep scour pool adjacent to the toe of the bank.

Bank instability and erosion may be evidence of the natural lateral movement of a stream as, for example, in meander migration. Although such an occurrence may not indicate a system that is out of sediment or hydrologic balance, lateral movement is problematic for culvert installations. Therefore, bank instability and erosion is not considered to be in dynamic equilibrium for the purposes of this manual.

In this step, a qualitative assessment of dynamic equilibrium is performed so the designer has a reasonable degree of assurance that an unstable stream reach will not prematurely threaten the culvert installation. The following discussion provides an assessment overview. HEC 20 (Lagasse, et al., 2001a) should be consulted for more detail.

Assessment of dynamic equilibrium in this step is a qualitative comparison of sediment transport potential through the project reach with the incoming sediment load at the upstream end of the project reach. If the supply is greater than the transport potential, aggradation will occur. If supply is less than the transport potential, degradation will occur. The assessment involves three components:

- Watershed reconnaissance for changes in supply.

- Project reach sediment transport assessment.

- Field observations of the project reach.

The watershed reconnaissance seeks to determine if changes in sediment supply to the project reach are occurring or likely to occur as a result of watershed changes. Undeveloped watersheds that are forecast to remain undeveloped suggest that sediment supply will not upset the potential for dynamic equilibrium, though natural hillside evolution could cause landslides, debris flows, and other forms of mass wasting. However, the designer should look for evidence of watershed changes that may increase sediment supply, such as recent forest fire activity, and for changes that may decrease sediment supply, such as significant urbanization or installation of a major dam and reservoir. The watershed reconnaissance should use sources available through office research as well as a field visit, if possible.

The project reach sediment transport assessment may be conducted using the observations of Lane (1955) relating channel hydraulics, sediment characteristics, and sediment transport as follows:

$$QS \propto Q_s D_{50}$$ (7.1)

where,

Q = flow discharge
S = energy slope
Q_s = sediment discharge
D_{50} = median sediment size

Lane's statement of proportionality is useful for qualitatively evaluating changes in sediment transport capability with changes in other parameters. His relationship may be rewritten as:

$$Q_s \propto \frac{QS}{D_{50}}$$ (7.2)

Equation 7.2 suggests that if the quantity on the right hand side of the equation is relatively constant through the project reach, the sediment transport capability is consistent through the reach. Therefore, the sediment transport assessment involves evaluation of the project reach to identify any imbalances in the three variables on the right hand side of Equation 7.2. However, Equation 7.2 is presented to provide a qualitative framework for assessment; a quantitative comparison is not recommended.

Finally, the watershed reconnaissance and transport assessment are reviewed in the context of field observations of the project reach. Are field observations of aggradation or degradation in the stream consistent with potential changes in sediment supply or changes in transport throughout the reach? If sediment supply does not appear to be affected by external drivers, sediment transport capability through the project reach is reasonably consistent, and there is no field indication of significant aggradation or degradation, the assessment indicates that the stream is in dynamic equilibrium. In this case, the designer proceeds to Step 5.

If dynamic equilibrium is not indicated, or the results of the assessment are ambiguous, the designer should investigate the sediment balance within in a more detailed framework. This is initiated by proceeding to Step 4.

7.4   STEP 4. ANALYZE AND MITIGATE CHANNEL INSTABILITY.

If the designer reaches this step, stream instability has either been identified or suspected. If instability is confirmed, it is prudent to determine the causes of the instability and develop an approach for mitigating the instability prior to proceeding with the crossing design.

Analysis and mitigation of stream stability is a complex endeavor that often requires recruitment of a specialist and is beyond the scope of this design procedure. Two references are cited here for further information. HEC 20 (Lagasse, et al., 2001a) may be consulted for the identification and diagnosis of stream instability problems. It cites three levels of analysis: 1) qualitative geomorphic analysis, 2) basic engineering analyses, and 3) mathematical and physical model studies. All assessments should begin with the qualitative geomorphic analysis and proceed to more complex analyses, as needed.

Once the causes of the stream instability are determined, HEC 23 (Lagasse, et al. 2001b) is a useful reference for design of measures to mitigate stream instability problems. A variety of measures are provided, including guidelines for design and construction of each measure. Other approaches may be employed. Another useful reference is McCullah and Gray (2005) "Environmentally Sensitive Channel and Bank Protection Measures." Other references may also be consulted. Regardless of the approach employed, mitigation measures must also support successful AOP.

When the analysis and mitigation activities are completed, the designer must determine if a culvert remains an appropriate design alternative for the location. If so, the designer proceeds to Step 5. If not, the culvert design procedure is terminated.

7.5 STEP 5. ALIGN AND SIZE CULVERT FOR Q_P.

The initial placement and sizing (and subsequent adjustments) of an embedded or open-bottom culvert for the peak design flow, Q_p, is determined in this step. The vertical and horizontal alignment, embedment depth, determination of Manning's n, consideration of debris, and evaluation of constructability are necessary considerations. Also, the shape, material, and inlet and outlet configurations affect culvert performance (Normann, et al., 2005).

As a starting point, the culvert should be designed to satisfy the customary hydraulic design criteria for the site. These criteria may include, but are not limited to, maximum allowable headwater, headwater depth to culvert rise (HW/D) ratio, minimum freeboard, and avoidance of overtopping. For an embedded culvert the culvert rise dimension, D, is vertical rise from the culvert crown to the bed, not the culvert invert. Subsequent steps in this design procedure may dictate changes to a culvert design produced solely based on hydraulic criteria.

In most situations, a single barrel culvert will be appropriate. Multiple barrel culverts on the same alignment are not preferred because they subdivide the culvert opening, increasing the potential for capturing debris, scour, differential aggradation/degradation between the barrels, and creation of artificial depth and velocity barriers. Any of these factors could reduce the hydraulic capacity and the capability for aquatic organism passage.

A floodplain relief culvert, or any culvert located at a higher elevation than the primary culvert, may be used to reduce the hydraulic burden on the primary (passage) culvert and facilitates floodplain hydrologic connectivity. The location and sizing of floodplain relief culverts would be considered prior to initiating this procedure. The design discharge passed by relief culverts should be deducted from the discharge passing through the primary stream culverts.

7.5.1 Vertical and Horizontal Alignment

Vertical and horizontal culvert alignment is addressed in HDS 5 (Normann, et al., 2005). Any culvert should be horizontally and vertically aligned with the existing channel bed to minimize disruption to the stream and to minimize costs associated with structural excavation and channel work. Consideration of AOP adds to the importance of such an alignment. A properly aligned culvert will also tend to reduce maintenance costs.

The vertical alignment should not exceed the slope of the streambed (or desirable streambed) and should fall within the range of slopes observed within the project reach. Vertical alignment may be one the parameters adjusted in later design steps, but should remain within the range of what is appropriate in the project reach.

Horizontal alignment must consider stream sinuosity and the relationship of the stream channel relative to the road. Most culverts are straight and sometimes cannot accommodate a sinuous channel. In other situations, the skew of the road versus the stream channel may result in an inordinately long culvert, which is not advantageous for AOP or cost.

Skewed culverts may create isolated zones of acceleration that impair passage. The extent of skew that may be problematic varies depending on slope and velocity, but skews of 15 degrees (0.26 radians) and larger may create undesirable flow and erosion conditions. Skewed culverts may also cause undesirable depositional problems. If the culvert cannot be aligned with the stream, special consideration must also be given to hydraulic conditions at the inlet and outlet so they do not create barriers for passage.

7.5.2 Length

Culvert length should be minimized to the extent feasible. Any culvert will exhibit characteristics that are unlike the natural channel. "Long" culverts present organisms seeking passage with a greater length to traverse than "short" culverts. Culverts also tend to reduce the sinuosity of a stream, which in turn increases the slope of the stream and ultimately stream velocity and erosion potential. (Sinuous culvert installations have been designed and installed.). The definition of "long" varies, in part, on the sinuosity of the stream.

7.5.3 Embedment

Embedment is intended to encourage AOP and to allow for downward adjustment in the bed within the culvert while still maintaining a natural bed. Conversely, the initial embedment level should allow for an upward adjustment to the bed while maintaining sufficient conveyance. Embedment provides the following benefits:

1. Flexibility to provide for vertical adjustments in the profile over the life of the culvert.

2. Sufficient bed thickness to allow for natural bed transformation processes, such as armoring.

3. Adequate wetted perimeter with increased bed roughness to mitigate acceleration of flows in the culvert.

For each of these benefits, increasing the depth of embedment enhances the ability of the culvert to provide the stated characteristics. However, once the desired function is obtained further increasing embedment may have undesirable consequences such as increasing the potential for subsurface flows and reducing the hydraulic capacity of the culvert such that Q_P requirements cannot be met.

Various embedment depth criteria have been developed by numerous organizations based on experience with the potential vertical adjustment of many types of streams (See Appendix E). For this procedure, the recommended embedment depth should be taken as the maximum of the following quantities:

1. Percent of culvert rise.

    a. 20 percent for box and pipe arch culverts.

    b. 30 percent for circular and elliptical culverts.

2. Multiple of natural bed material D_{95}.

    a. One times the D_{95} for box and pipe arch culverts.

    b. Two times the D_{95} for circular and elliptical culverts.

3. 2 ft (0.61 m).

These embedment criteria capture a wide range culvert sizes, shapes, and slopes, and assume significant channel degradation or scour is not expected to occur over the life of the structure. If a wider range of vertical adjustment is anticipated, the embedment should be increased, which may, in turn, require use of a larger culvert or alternative culvert type.

7.5.4 Bed Gradation

The bed gradation within the culvert should match the bed gradation within the streambed as closely as possible. To this end, the bed material should be sampled in the project reach where the bed slope is similar to that proposed for the culvert. Critical features of the bed gradation within the culvert are (FSSWG, 2008):

1. Large particles (D_{95}, D_{84}, and D_{50}) should be properly sized to provide bed structure and buttress finer material.

2. The entire bed mix should be well graded (poorly sorted). A dense, stable bed requires all particle sizes, so no gaps should exist between any classes of material in the design bed mix.

3. The percentage of smaller fractions (sand, silt, and clay) should approximate the adjacent reach, but should also be adequate to limit bed interstitial flow. The D_5 fraction should be no larger than 0.079 in (2 mm).

When designing a well-graded bed within the culvert adjacent to an armored streambed, the designer must consider the potential for lowering of the bed in the culvert as it goes through the armoring process. As smaller particles are washed out, the bed will lower. Unless, new material can be expected from the upstream reach or the bed design uses the armored gradation, the design bed should be increased in depth to account for the drop expected to occur when the stream becomes armored. Appendix F summarizes selected procedures for designing bed gradations.

7.5.5 Manning's n

Manning's n values must be estimated for the culvert material and the bed material in the culvert. In addition, a composite n value within the embedded culvert must be calculated. See Appendix C for a selection of recommended methodologies. This appendix is not intended to be exclusive; the designer may use other methods appropriate for the situation.

7.5.6 Debris

According to Normann, et al. (2005), debris is defined as any material moved by a flowing stream. Debris includes some combination of floating material, suspended sediment, and bedload. A stream's propensity for carrying debris is based upon watershed land uses and certain stream and floodplain characteristics, such as:

1. Stream velocity, slope, and alignment.

2. Presence of shrubs and trees on eroding banks.

3. Watershed land uses, particularly logging, cultivation, and construction.

4. Stream susceptibility to flash flooding.

5. Storage of debris and materials within the flood plain (logs, lumber, solid waste, etc.)

6. Recent occurrence of fires.

Woody debris may have a positive effect on fish if it becomes snagged or otherwise trapped in the stream channel thereby creating habitat. Such debris may also cause changes in a stream cross-section or profile depending on the size and type of debris.

Debris may also be trapped at a culvert inlet or within the culvert barrel. This is undesirable from a hydraulic perspective. Trapped debris will reduce the capacity of the culvert forcing water to back up or escape via an alternate route, potentially causing damage to the roadway embankment and/or adjacent properties. Debris may also contribute to scour and erosion problems. Debris trapped at a culvert inlet or outlet may create a barrier to fish passage.

The production and transport of debris is complex and beyond the scope of this document. However, the size and shape of the culvert opening, when operating during the flood peak, govern the ability of the culvert to pass the debris. Minimizing sharp edges at inlets, sills, and baffles, if present, also reduces the chance of trapping debris. All debris cannot be passed through a culvert, just as all debris is not passed in the natural channel. A reasonable accommodation for debris can be made, however. Hydraulic Engineering Circular Number 9 (Bradley, et al., 2005) should be consulted for further information.

In traditional culvert design for flood flows, debris is addressed via the headwater depth to culvert rise (HW/D) ratio. If debris is not considered a concern at a given location, the HW/D ratio traditionally may range from 1.0 to 1.5 depending on the design event and local policies. If a site assessment reveals that debris is a concern, this criterion is often lowered to 0.8 or less to provide space for debris to pass. The analyses of Steps 6, 8, and 9 will usually lower the HW/D ratio below these traditional values in order to achieve the stability of the streambed material and facilitate AOP.

Bridge design addresses debris by the use of freeboard and by insuring that there is a sufficient width of opening at the required freeboard to pass debris. For either culvert or bridge design, it is necessary that the size and shape of the opening accommodate the size and shape of debris for the debris to pass.

7.5.7 Culvert Analysis and Design Tools

Two primary culvert analysis tools are available for applying this procedure: HY-8 and HEC-RAS. Other appropriate culvert simulation programs may also be used. Whichever tool is applied, it must be able to simulate inlet and outlet control for a wide range of flow conditions, embedment, and the differences in hydraulic roughness between the bed and culvert material.

The HY-8 computer program was originally developed to analyze culvert hydraulics during design flood events. This focus on high-flow capacity does not require detailed analysis of very low-flow hydraulics. Consequently, some HY-8 results for very low flows in culverts are approximations. The most precise reporting of the inlet and outlet conditions is found with the "water surface profile data" so these data should be retrieved and used in the design. In some cases, these values may be slightly different than values reported in the "Culvert Summary Table."

When using HEC-RAS for design, the inlet and outlet water surface elevations are provided in the detailed culvert output. With the depths, the velocities are calculated from the continuity equation and the energy slope is estimated from Manning's equation assuming uniform flow. Shear stresses should not be read directly from the HEC-RAS output because these values represent average, rather than maximum, shear stresses in a cross-section. When using HEC-RAS with embedded culverts, the designer must remember to subtract the embedment depth from reported depths in the culvert.

HEC-RAS offers a "LID" function that provides an alternative modeling technique for culverts. Effectively, the LID function would give a water surface profile through the culvert, which is not available when modeling the culvert directly as a culvert in HEC-RAS. The LID feature would also provide some flexibility in describing the geometry of the bed. However, one would also sacrifice the inlet and outlet energy loss calculations unique to culverts requiring detailed consideration of the appropriate energy loss calculations with the LID option. The appropriate choice of method depends on the situation and the skill of the modeler.

On completion of this step, a culvert shape, size, material, alignment, and embedment have been determined.

7.6 STEP 6. CHECK CULVERT BED STABILITY AT Q_H.

Characteristics of the streambed material within the culvert are established in Step 5. In this step, the permissible and applied shear stresses (or critical and actual unit discharges) are estimated and compared to determine if the streambed material within the culvert is stable at Q_H.

The methods described for evaluating stability are the modified Shield's method and the critical unit discharge method. In most situations, the modified Shield's method can be applied for slopes up to 5 percent and the critical unit discharge method for slopes from 3 to 10 percent. In general, the modified shear stress method is recommended for slopes from 0 to 3 percent and the critical unit discharge approach is recommended for slopes from 5 to 10 percent. Between 3 and 5 percent, both methods should be applied taking the most conservative approach for design. For 10 to 20 percent slopes, the critical discharge method is applicable for uniform bed materials, but has not been tested for non-uniform materials.

The designer may select other methods provided reasonable justification and documentation is given. Appendix D may be consulted for more information.

If the streambed material in the culvert is determined to be stable by application of an appropriate method, the designer can bypass Step 7 and move to Step 8. If not, the relative mobility of the streambed material in the culvert and in the stream must be checked in Step 7.

7.6.1 Permissible Shear Stress

The permissible shear stress approach is based on a comparison of the applied shear stress to the ability of a bed material to resist movement, that is, its permissible shear.

7.6.1.1 Noncohesive Materials

Values for permissible shear stress for a wide range of sizes greater than 50 mm (2 in) are based on research conducted at laboratory facilities and in the field. For more uniformly graded materials, permissible shear stress is calculated based on a characteristic grain size from the following equation:

$$\tau_p = F_* (\gamma_s - \gamma) D_{50} \tag{7.3}$$

where,

τ_p	=	permissible shear stress, lb/ft^2 (N/m^2)
F_*	=	Shield's parameter, dimensionless
γ_s	=	specific weight of the stone, lb/ft^3 (N/m^3)
γ	=	specific weight of the water, 62.4 lb/ft^3 (9810 N/m^3)
D_{50}	=	stone size for which 50 percent, by weight of the bed is smaller, ft (m)

Typically, a specific weight of stone of 156 to 165 lb/ft^3 (24,500 to 25,900 N/m^3) is used, but the site-specific value should be used.

Shield's parameter is expressed as a function of Reynolds number as shown in Table 7.1.

Table 7.1.   Selection of Shields' Parameter.

Reynolds number	F*
≤ 4x10^4	0.047
4x10^4<R$_e$<2x10^5	Linear interpolation
≥ 2x10^5	0.10

From Kilgore and Cotton (2005)

The particle Reynolds number is defined, based on the characteristic grain size, as:

$$R_e = \frac{V_* D_{50}}{\nu}$$
(7.4)

where,

   R_e = particle Reynolds number, dimensionless
   V_* = shear velocity, ft/s (m/s)
   ν = kinematic viscosity, 1.217x10^{-5} ft^2/s at 60 deg F (1.131x10^{-6} m^2/s at 15.5 deg C)

Shear velocity is defined as:

$$V_* = \sqrt{gyS}$$
(7.5)

where,

   g   = gravitational acceleration, 32.2 ft/s^2 (9.81 m/s^2)
   y   = maximum channel depth, ft (m)
   S   = channel slope, ft/ft (m/m)

Equations 7.3, 7.4, and 7.5 are valid for slopes up to 10 percent.

Most natural bed materials cannot be considered uniformly graded so that direct application in such cases is not valid. The interaction between stone sizes larger and smaller than the D_{50} has an effect on the bed stability. Therefore, the modified Shield's equation is used to capture this nonuniformity in natural bed materials based on the D_{84} and D_{50} of bed material:

$$\tau_p = F_* (\gamma_s - \gamma) D_{84}^{0.3} \, D_{50}^{0.7}$$
(7.6)

where,

τ_p = permissible shear stress, lb/ft^2 (N/m^2)

F$_*$ = Shields parameter for D$_{50}$ particle size (this value is obtained from Table 7.1)

γ_s = specific weight of the stone, lb/ft^3 (N/m^3)

γ = specific weight of the water, 62.4 lb/ft^3 (9810 N/m^3)

D$_{50}$ = stone size for which 50 percent, by weight of the bed is smaller, ft (m)

D$_{84}$ = stone size for which 84 percent, by weight of the bed is smaller, ft (m)

As discussed in Appendix D, Equation 7.6 has only been confirmed for applications with a bed slope of 5 percent or less, D$_{84}$ less than or equal to 9.8 in (250 mm), and a D$_{84}$/D$_{50}$ ratio less than or equal to 30.

For fine-grained, noncohesive soils (D$_{75}$ < 0.05 in (1.3 mm)) permissible shear stress is relatively constant and is conservatively estimated at 0.02 lb/ft^2 (1.0 N/m^2). For coarse grained, non-cohesive soils (0.05 in (1.3 mm) < D$_{75}$ < 2 in (50 mm)) the following equation applies.

$$\tau_{p,soil} = \alpha D_{75} \tag{7.7}$$

where,

$\tau_{p,soil}$ = permissible soil shear stress, lb/ft^2 (N/m^2)

D$_{75}$ = particle size where 75 percent of the material is finer, in (mm)

α = unit conversion constant, 0.4 (CU), 0.75 (SI)

7.6.1.2 Cohesive Materials

Cohesive soils are largely fine grained and their permissible shear stress depends on cohesive strength and soil density. Cohesive strength is associated with the plasticity index (PI), which is the difference between the liquid and plastic limits of the soil. The soil density is a function of the void ratio (e). The basic formula for permissible shear on cohesive soils is the following.

$$\tau_{p,soil} = \left(c_1 PI^2 + c_2 PI + c_3\right)\left(c_4 + c_5 e\right)^2 c_6 \tag{7.8}$$

where,

$\tau_{p,soil}$ = soil permissible shear stress, lb/ft^2 (N/m^2)

PI = plasticity index

e = void ratio

$c_1, c_2, c_3, c_4, c_5, c_6$ = coefficients (See Appendix D)

A simplified procedure for determining permissible shear for cohesive soils is provided in Appendix D.

The method given by Equation 7.8 does not take into account three important factors: 1) the presence of coarse material (to have cohesive properties, only 10 percent of the mixture must be clay sized) and its effect on the breakdown and ultimate initiation of motion of cohesive flocs; 2) the consolidation history of the bed layer(s); and 3) and the water column sediment concentration. However, the affects of these factors are complex to predict and are the subject of ongoing research.

7.6.1.3 Applied Shear Stress

Hydraulic analyses of the culvert are performed to determine the shear stresses experienced by the bed material in the culvert. The hydraulic analyses provide estimates of depth, energy slope, and velocity at each location. The applied shear stress for the channel and the culvert can be retrieved from the hydraulic model or computed using:

$$\tau_d = \gamma y S \tag{7.9}$$

where,

τ_d = maximum applied shear stress, lb/ft^2 (N/m^2)

γ = unit weight of water, lb/ft^3 (N/m^3)

y = maximum depth, ft (m)

S = energy slope, ft/ft (m/m)

Flow within the culvert is generally not uniform and is, therefore, not characterized by a single hydraulic radius and energy slope. In outlet control, the conditions at the inlet and outlet will bracket the range of hydraulic conditions within the culvert. Therefore, shear stress at both the inlet and outlet should be computed to determine the most critical (high stress) location. For inlet control with low tailwater, the conditions at the inlet and outlet will also bracket the range of hydraulic conditions within the culvert. However, for inlet control with high tailwater, a shallower flow condition may occur within the culvert barrel rather than at the inlet or outlet. The latter will occur for inlet control conditions where the outlet depth is determined by the tailwater depth or full flow conditions.

If such a condition is noted, a third shear stress should be computed using the minimum depth in the culvert flow profile. If the full water surface profile within the culvert is not available, assuming the flow in the culvert reaches normal depth before the tailwater begins to influence the profile is conservative.

If the applied shear stress is less than or equal to the permissible shear stress, the designer proceeds to Step 8. If not, the designer moves to Step 7.

7.6.2 Critical Unit Discharge

In situations where depth is difficult to measure and/or define because the size of the roughness elements relative to the depth is high, computation of unit discharge and the critical unit discharge provides an alternative measure of stability.

Unit discharge is defined as the discharge above the active channel bed divided by the width of the active channel bed:

$$q = Q_a / w_a \tag{7.10}$$

where,

q = unit discharge, ft^3/s/ft (m^3/s/m)

Q_a = active channel discharge, ft^3 (m^3)

w_a = active channel bed width, ft (m)

Within a culvert, there is no practical distinction between active channel and other channel components such as a floodplain. Therefore, the active channel discharge is the total discharge and the active channel width is the flow top width. If the water surface elevation in the culvert is above the springline of the culvert and flow top width is less than the culvert span, the culvert span should be taken as the active channel bed width.

For fairly uniform materials, the critical discharge is computed based on the characteristic grain size:

$$q_{c-D50} = \frac{0.15g^{0.5}D_{50}^{1.5}}{S^{1.12}}$$ (7.11)

where,

q_{c-D50} = critical unit discharge to entrain the D_{50} particle size, ft³/s/ft (m³/s/m)
D_{50} = median or 50th percentile particle size, ft (m)
g = gravitational acceleration, 32.2 ft/s² (9.8 m/s²)
S = bed slope, ft/ft (m/m)

To adapt to the more typically nonuniform bed materials found in a natural channel, the critical unit discharge for entraining the D_{84} particle size is determined by:

$$q_{c-D84} = q_{c-D50}(D_{84}/D_{50})^b$$ (7.12)

where,

q_{c-D84} = critical unit discharge to entrain the D_{84} particle size, ft³/s/ft (m³/s/m)
D_{84} = 84th percentile particle size, in (mm)
D_{50} = median or 50th percentile particle size, in (mm)

The exponent b is a measure of the range of particle sizes that make up the channel bed. It quantifies the effects on particle entrainment of smaller particles being hidden and of larger particles being exposed to flow. Calculate the exponent from:

$$b = 1.5(D_{16}/D_{84})$$ (7.13)

where,

D_{84} = 84th percentile particle size, in (mm)
D_{16} = 16th percentile particle size, in (mm)

As discussed in Appendix D, Equations 7.12 and 7.13 were derived from limited data and are most appropriate for conditions summarized in Table 7.2.

Table 7.2.   Parameter Ranges for Critical Unit Discharge for D_{84}.

Parameter	Low	High
Slope (%)	3.6	5.2
Width, ft (m)	20 (6.1)	36 (11)
D_{16}, in (mm)	1.3 (32)	2.3 (58)
D_{50}, in (mm)	2.5 (72)	5.5 (140)
D_{84}, in (mm)	6 (156)	10 (250)

If the unit discharge is less than or equal to the critical unit discharge, the designer proceeds to Step 8. If not, the designer moves to Step 7.

7.7   STEP 7. CHECK CHANNEL BED MOBILITY AT Q_H.

The objective of this step is to evaluate whether the culvert design should be altered to achieve a stable bed within the culvert during the high passage discharge, Q_H. The assessment compares the relative mobility of the streambed material in the culvert to the relative mobility of the streambed material in the upstream and downstream channel using shear stress or unit discharge, as appropriate.

Bed behavior varies with channel type. Noncohesive channels range from dune-ripple channels with sand to gravel beds to cascade channels with large boulder beds. Other channels have either bedrock or cohesive soil beds. Bed behavior is also influenced by rock clusters, vegetation, and bedrock outcrops that provide diversity to the natural channel. The designer should be aware of these differences in stream channels to assess bed mobility.

Conditions within the culvert are estimated in Step 6. If HY-8 is used to size the culvert in Step 5 it should be rerun with the high passage flow, Q_H. The stream cross-sections are analyzed by a series of uniform depth computations.

If HEC-RAS, or equivalent, is used to size the culvert, it should be rerun using the high passage flow, Q_H. HEC-RAS provides the hydraulic values to compute the shear stress or unit discharge at each cross-section. (Shear stress is reported by HEC-RAS in the channel cross-sections, but these values should not be used because they represent the average, not maximum, shear stress in the cross-section.)

Interpretation of the shear stresses, or unit discharge, should conform to the following guidance:

1. If the maximum applied shear stress, τ_d, or unit discharge for any channel cross-section is less than the permissible shear stress or critical unit discharge, respectively, then a redesign of the culvert to achieve a stable bed should be undertaken. Return to Step 5.

2. If the maximum applied shear stress, τ_d, or unit discharge for all channel cross-sections is greater than the permissible shear stress or, the bed may be considered mobile.

    a. If the culvert shear stresses or unit discharges fall within the range observed in the channel the culvert is adequately sized for the purposes of this step. Go to Step 8.

    b. If the culvert shear stresses or unit discharges exceed the range observed in the channel the culvert is not adequately sized. Return to Step 5.

7.8   STEP 8. CHECK CULVERT BED STABILITY AT Q_P.

To insure the long-term performance of the embedded material within the culvert, the stability of the bed material is assessed to determine if it is stable at Q_P. In many situations, the streambed material in the culvert and the upstream and downstream channel will be mobile when subjected to significant flow rates such as the design flood peak, Q_p. Depending on the capacity of the stream for replenishment, streambed material may be redeposited within the culvert on the receding limb of the flood hydrograph or during subsequent smaller runoff events resulting in a sustainable bed within the culvert.

However, to be able to demonstrate that replenishment at a given site is likely requires significant sediment transport analyses that are beyond the scope of this design procedure. Therefore, this procedure relies on a worst-case assumption of no replenishment and the stability of a designed sublayer, if needed, to enhance the re-establishment of the natural channel substrate.

7.8.1 Bed Stability

An applied shear stress within the culvert for Q_p is estimated. If this shear stress is less than or equal to the permissible shear stress for the bed material, the culvert bed is stable and the designer should proceed to Step 10. If it is not stable, proceed to Step 9 to design a stable sublayer. Stability of the native streambed material at the design peak flood is unlikely except for relatively large bed material.

7.8.2 Pressure Flow

At the design flood peak, it is possible that all or part of the culvert is flowing under pressure, i.e. the barrel is full and a free water surface does not exist for that portion of the culvert. This condition will not be experienced when the headwater depth to rise ratio is less than one and may not be experienced in many other situations. Since assessing bed stability and replenishment requires computation of shear stresses, a method for computing shear stress under pressure conditions is needed.

Theoretically, permissible shear stress for the culvert bed material is the same whether or not the culvert is flowing under pressure. The resistance to motion is a function of the bed material properties. However, pressure flow conditions require an adaptation for computation of the applied shear stress since the culvert confines the water surface. The applied shear stress for a conduit under pressure is based on the hydraulic grade line, rather than the free surface depth. The appropriate "depth" for calculating the applied shear stress under pressure flow conditions is the height of the energy grade line above the bed minus the velocity head:

$$y = EGL - INV - \frac{V^2}{2g} \tag{7.14}$$

where,

y	=	"depth," ft (m)
EGL	=	energy grade line elevation at point of analysis, ft (m)
INV	=	bed elevation at point of analysis, ft (m)
V	=	velocity (Q/A), ft/s (m/s)
g	=	acceleration due to gravity, ft/s^2 (m/s^2)

7.9 STEP 9. DESIGN STABLE BED FOR Q_P.

If the streambed material within the culvert is unstable at Q_P, a stable sublayer is recommended to maintain the bed profile in the culvert during the peak design flow, maintain roughness in the bottom of the culvert barrel, and promote deposition of native material. This recommendation presumes that at flood flows the shear stresses within the culvert will be greater than those in the upstream and downstream channels because the width in the culvert is limited and there may be no floodplain relief access.

In some channel systems, the bed within the culvert may be naturally replenished. If the designer has performed a site-specific sediment transport analysis documenting this to be the case, a stable sublayer may not be needed.

An oversized sublayer beneath the native streambed material is the primary tool currently available to achieve a stable bed within the culvert. The layer of native streambed material will overlay the oversized sublayer and will continue to provide the desired substrate and the potential for vertical adjustment. However, the thickness of the native streambed material layer should be adjusted recognizing the presence of the oversized sublayer. With this in mind, the minimum native streambed material layer should be the maximum of the following:

1. One times the D_{95} for all culvert types.

2. 1 ft (0.3 m).

The minimum oversized sublayer material thickness should be:

1. 1 times the D_{95} for box and pipe arch culverts

2. 1.5 times the D_{95} for circular and elliptical culverts.

The minimum oversized sublayer thickness is based on the need to establish a layer of material uniformly over the bottom of the culvert. The added thickness for circular and elliptical culverts is to allow the larger size fractions to be placed throughout the embedment including near the culvert sides.

The minimum native streambed material layer provides for vertical adjustment within the culvert. In combination with the oversized layer, the native streambed material layer provides the overall functions of the embedment criteria described in Section 7.5.3.

The sum of the two layers should be compared to the minimum embedment that would be required using the criteria from Step 5 for a single native streambed material layer. If the sum of the two layers is less than the single layer embedment, the native bed material layer should be increased until the combined layer thickness is equal to the single layer embedment thickness.

Depending on the full gradation of the oversize layer, especially the smaller fractions, the designer may consider an end sill to prevent interstitial flow through the bed, but they should not be relied on to retain native bed materials.

The use of baffles or sills may provide some benefits for retaining the bed material within the culvert or promoting redeposition of native streambed materials after flood events. However, these benefits have not been demonstrated and are, therefore, not recommended for this purpose. (See Appendix G for a discussion of baffles and sills.)

7.9.1 Oversized Bed Material Gradation

Lagasse, et al. (2006) provide guidance for determining size fractions from the D_{10} through the D_{100} based on the D_{50} for rock layers designed to resist shear stresses. Simplifying their recommendations for the larger size fractions results in the following equations:

$$D_{16} = 0.7D_{50} \tag{7.15a}$$

$$D_{84} = 1.4D_{50} \tag{7.15b}$$

$$D_{95} = 1.9D_{50} \tag{7.15c}$$

Several approaches are available for determining the smaller size fractions, D_{16} and D_5. As discussed previously, the D_5 should be no larger than 0.079 in (2 mm) to limit interstitial flow through the sublayer. D_{16} may be calculated based on Equation 7.15a, the Fuller Thompson method based on the D_{50} (Appendix F), the native bed material D_{50}/D_{16} ratio, or by visually completing the gradation curve given the larger size fractions and D_5.

However, use of Equation 7.15a results in a relatively uniform gradation, which may make it difficult to include smaller fractions effectively. By comparison, use of the Fuller Thompson equation (Appendix F) with the parameter, m, equal to 0.5 results in an estimate of D_{16} equal to ten percent of D_{50}. The final determination should consider the site, the native gradation, and interstitial flow.

The ability to specify the resulting mix in a construction contract is important in the final selection. Limits on oversizing of bed material include availability of the larger material, constructability using the larger (heavier) material, and effects on embedment depth.

7.9.2 Design Equations

For slopes less than 5 percent, the modified shear stress approach may be used. Substituting Equation 7.15b into Equation 7.6 yields the permissible shear stress as:

$$\tau_p = 1.1\,F_*\left(\gamma_s - \gamma\right)D_{50} \tag{7.16}$$

where,

τ_p = permissible shear stress, lb/ft^2 (N/m^2)

F_* = Shields parameter for D_{50} particle size (this value is obtained from Table 7.1)

γ_s = specific weight of the stone, lb/ft^3 (N/m^3)

γ = specific weight of the water, 62.4 lb/ft^3 (9810 N/m^3)

D_{50} = stone size for which 50 percent, by weight of the bed is smaller, ft (m)

An initial trial D_{50} is selected based on the shear stresses experienced with the native material, recognizing that with a larger D_{50}, roughness will increase. As a result, depth and shear stress will also increase.

For slopes between 3 and 10 percent, the critical unit discharge approach may be used. Equation 7.11 is used to determine the critical unit discharge for the D_{50} stone size as in Step 6. Substituting Equation 7.15b into Equation 7.12 provides the following equation:

$$q_{c-D84} = q_{c-D50}\left(1.4\right)^b \tag{7.17}$$

where,

q_{c-D84} = critical unit discharge to entrain the D_{84} particle size, ft^3/ft (m^3/m)

q_{c-D50} = critical unit discharge to entrain the D_{50} particle size, ft^3/ft (m^3/m)

b = exponent derived from Equation 7.13

For slopes between 3 and 5 percent, both methods are applied and the largest D_{50} is selected for design.

7.9.3 Design Alternatives

The appropriate selection of a stable bed design will consider passage benefits, costs, and constructability. An alternative to designing a stable bed is to provide relief culverts to divert the peak design flow away from the passage culvert. By diverting the peak flows, the stress in the culvert is reduced. If the designer wishes to evaluate flow diversion to other culverts, return to Step 5.

If a stable bed is feasible by oversizing the bed material, the designer should recompute the Manning's n, composite roughness, and resulting hydraulics in the culvert to verify that the design criteria for the design flood peak are satisfied assuming the native bed material has been washed out by the time the peak of the hydrograph occurs. If so, continue to Step 10.

If a stable bed with oversize material is not feasible, or results in excessive embedment in the culvert, the designer should return to Step 5 to select a new culvert size, shape, material, and/or slope. The designer will make the assessment of whether embedment depths are excessive, but depths exceeding 40 or 50 percent of the culvert rise are likely to result in unsatisfactory reductions in capacity.

7.10 STEP 10. CHECK CULVERT VELOCITY AT Q_H.

A check of the velocity within the culvert is conducted to determine if velocities experienced within the culvert are within the range of those experienced upstream and downstream of the culvert within the project reach. The maximum velocity in the culvert should not exceed the maximum velocity in the channel cross-sections. Specific fish swimming abilities are not needed for this step since the step is a relative comparison between the culvert and the natural channel.

All velocities used in this step are cross-section averages. It is recognized that there are variations within a cross-section that result in lower velocities near the boundaries and higher velocities away from the boundaries. However, since variations occur in both the stream and within the culvert to some degree, detailed analysis of such variations has not been demonstrated to be justified.

In Step 6, the designer has determined the culvert inlet and outlet hydraulic variables, including velocity at Q_H. In some inlet control situations, these variables will have also been determined within the barrel of the culvert. From these computations, the highest velocity, and the length over which it occurs, is identified. The designer also determines the cross-section velocities in the channel cross-sections.

The designer is cautioned that if the controlling culvert velocity is critical velocity at the outlet, the analysis (HY-8, HEC-RAS, or other approach) should be reviewed to verify that this is a reasonable condition. Often, it indicates a poor vertical alignment between the culvert outlet and the downstream cross-section.

Since velocity is important for passage not only for its absolute value, but also for the length of channel over which it applies, the designer must compare the length of channel section to the culvert length to verify that the most severe conditions do not occur in the culvert.

It may be desirable to add velocity diversity within a culvert to mimic that observed in the stream. This may be accomplished by placing larger rocks in a pattern similar to that found in the native stream within the culvert barrel. This may be of particular value in longer culverts. It is recommended that the larger rocks be stable at the peak design flow, Q_p, since natural replacement of these rocks within the barrel is unlikely. After developing a design with such rocks, the designer should reevaluate culvert capacity considering the changes in roughness and cross-sectional area contributed by their placement. Clogging potential for debris to be trapped on the larger rocks should also be assessed.

If the culvert velocity exceeds the stream cross-section velocities, considering the length of culvert/channel over which they occur, the designer should go back to Step 5 and revise the culvert size, slope, or both. If an oversize bed sublayer has been incorporated in the culvert design in Step 9, the designer should be aware that increasing the culvert size or lowering the culvert slope may eliminate the need for the oversize sublayer.

If the culvert velocity is satisfactory, proceed to Step 11.

7.11 STEP 11. CHECK CULVERT WATER DEPTH AT Q_L.

A check of the water depth in the culvert at the low passage flow, Q_L, is conducted. Hydraulic analyses are repeated at Q_L to determine the thalweg flow depths in the channel cross-sections and within the culvert.

The limiting flow depth will be at the inlet or outlet of the culvert except in those inlet control situations where it will occur in the barrel of the culvert. The smallest depth is taken as the limiting culvert flow depth for passage. The thalweg flow depths are also retrieved from each channel cross-section. The smallest thalweg depth in the channel cross-sections is taken as the limiting stream cross-section flow depth. If the limiting culvert flow depth is greater than or equal to the limiting stream cross-section flow depth, the water depth in the culvert is satisfactory.

Some streambed materials are capable of conveying flow within the bed. At the low passage flow, this interstitial flow may be significant resulting in an overestimate of depths in the stream and within the culvert. If the bed material in the channel and the culvert are essentially the same, the check noted above will be valid since any interstitial flow will affect both the channel and the culvert. However, if the void ratio within the culvert bed is larger than in the streambed or if the rock size in the culvert bed is larger than in the streambed, interstitial flow potential should be assessed. If appropriate, baffles below the bed level may be used to cut off interstitial flow. Appropriate use of fines in the bed material to reduce interstitial flow is discussed in Appendix F.

If the water depth in the culvert is satisfactory, skip to Step 13. Otherwise, further adjustments to the culvert bed design are necessary. See Step 12.

7.12 STEP 12. PROVIDE LOW-FLOW CHANNEL IN CULVERT.

Provide a low-flow or pilot channel through the culvert that provides an equivalent maximum depth to that required for the low passage flow as determined in Step 11. The low-flow channel should be designed such that it does not reduce the flood flow capacity. Other than providing for the required depth, the dimensions are not critical. A triangular shape with 1:5 side slopes (or flatter) is a good starting point (FSSWG, 2008).

For designs where the bed material in the culvert is identical to the bed material in the adjacent stream channels, a low-flow channel may form naturally. If so, the question becomes how long will this take to occur? This question is difficult to answer with any certainty, though mobile beds (silts, sands, and gravels) will react more quickly. If the designer is not confident with natural formation, the low-flow channel should be included in the initial installation.

7.13 STEP 13. REVIEW DESIGN.

The culvert design is completed. Review the design for its compatibility with the project objectives, environmental requirements, and construction and maintenance costs. Repeat design process at Step 5 with alternative culvert types and/or shapes, if desired.

If an open-bottom culvert is selected, foundation design must be conducted, including consideration of scour. The need for outlet protection to prevent excessive outlet scour should also be evaluated. If required, the protection type and design should be compatible with the AOP objective. These assessments are beyond the scope of this procedure.

This page intentionally left blank.

CHAPTER 8 - CONSTRUCTION

The following construction topics have unique applications in culverts designed for aquatic organism passage (AOP). Topics are not covered in-depth, however, links to pertinent references are included.

8.1 TIMING

Timing of in-stream work will need to correspond to specific periods allowable by resource agencies. An in-stream work permit will generally be required.

8.2 STREAM PROTECTION

Construction activities must be planned and executed so that negative impacts to the stream are minimized. Staging and materials storage areas should be located and managed to prevent untreated runoff from reaching the stream. Bank revegetation, where appropriate, should be carefully planned and executed. Like all construction projects, erosion and sediment control measures should be properly implemented.

8.3 CONSTRUCTABILITY

The successful construction of embedded culverts is contingent on the ability of crews to place bed material and larger roughness elements within the structure. In general, this leads to a practical minimum span of 6 ft (1.8 m) (Bates et al., 2003), although 5 ft (1.5 m) installations are reportedly placed routinely in Alaska (Gubernick, 2007). Depending on the size of the barrel and bed materials, placement may be accomplished by a number of methods including Dingo Loaders, rock chutes, wheelbarrows and trail building equipment. Open-bottom culverts and enclosed culverts that are installed in pieces, such as structural plate arches and cast in place concrete boxes, avoid such a limitation.

8.4 STREAMBED MATERIAL AND PLACMENT

Due to the difficulty involved with mixing bed materials on site, it is recommended that material be mixed prior to placement, except when backfilling large key elements with fines. Channel margin or bankline features, if needed, must be placed by hand (United States Forest Service, 2006a). There also may be practical limits on the availability of the oversize bed gradations specified. If so, some adjustment to oversize bed designs may be necessary.

Materials excavated for embedded culvert or open-bottom foundation installations should be reserved and replaced within the streambed if the material is suitable. Materials should not be dredged from adjacent reaches of the stream, which among other consequences, may initiate a headcut. Suitable material should be obtained from gravel pits or quarries or other offsite sources.

Quality assurance and quality control (QA/QC) are very important for all vertical controls, especially for the placement of all key pieces, channel margin or bankline features, and oversize bed materials. Provisions for QA/QC should be included as part of the construction contract.

8.4.1 Sealing Voids

In culverts with placed sediments, especially those involving the use of oversized sediment sublayers, it is important to limit permeability. Without such considerations, a significant portion of flow may seep through interstitial voids, causing the stream to flow below the bed for periods of time. Methods to limit permeability include placement of geotextile (Browning 1990) and an adequate proportion of fine sediments in bed mixes (Bates et al., 2003; FSSWG, 2008). During construction, fines can be power-washed into voids to ensure and expedite bed sealing. This washing procedure will also decrease the sediment concentration entering the stream system after the first flow event.

8.4.2 Compaction

For oversized bed installations, bed material is placed in thin layers with thickness appropriate for the slope and for the size of the mix, compacted, and covered with filler material to be washed into voids (United States Forest Service 2006a). Smaller material should be well compacted around larger elements (FSSWG, 2008).

CHAPTER 9 - POST CONSTRUCTION

Properly designed and constructed passage culverts will require inspection and monitoring to ensure continued performance, especially in the first few years, to evaluate the potential to collect debris or to scour/aggrade the streambed. Maintenance needs will be identified through inspection and monitoring.

9.1 STRUCTURAL INSPECTION

Culverts that qualify as bridges (total span exceeds 20 ft (6.1 m)) must be inspected every two years using 23 CFR 650 Subpart C of the National Bridge Inspection Standards as a guide (FHWA, 2004). This inspection includes checks of all underwater elements and checks for scour and fill at the crossing.

There are few, if any, documented schedules for culvert inspection and maintenance. Standard culvert problems and treatments are listed in the Federal Highway Administration *Culvert Repair Practices Manual Volume I* (Ballinger and Drake, 1995), and CALTRANS has supplemental guidelines for use in their transportation system (California Department of Transportation, 2006). Wyant (2002) provides an overview of existing practices for the assessment and rehabilitation of culverts.

Inspection is advisable at regular intervals and ideally during flood events. This may be especially important at installations with significant amounts of Large Woody Debris (LWD), or at crossings with a propensity to collect debris.

9.2 PASSAGE MONITORING

Although much research has been done to understand the requirements of aquatic organism passage, gaps in knowledge, nuances in behavior, and lack of adequate hydraulic and hydrological data contribute to a degree of uncertainty in the long-term passage performance of any given structure. A monitoring program will help ensure that the structure impact on aquatic organism passage is more clearly understood, allowing future criteria for assessment and design to be more effective (General Accounting Office, 2001).

Monitoring should begin with clear project goals that will allow the development of measurable parameters to allow "success" to be quantified (Committee on Restoration of Aquatic Ecosystems, 1992). Ideally, monitoring might include direct observation of fish movement and utilization, but should at least focus on project compliance with design specifications such as substrate retention and the ability to maintain passable conditions (Furniss, 2006).

Beginning with project goals in mind, parameters and field methods should be aimed at comparing current physical conditions to design performance criteria. Building upon this type of analysis, Harris (2005) developed criteria for fish passage installation effectiveness monitoring in California that is summarized in Table 9.1. The table includes a mix of qualitative and quantitative criteria and parameters. Resource agencies with jurisdiction or consultation responsibilities for a given site should be consulted to identify appropriate quantitative bioassessment protocols for addressing questions 3 and 4 in the table.

Table 9.1.   Monitoring Evaluation.

Monitoring Question	Effectiveness Criteria	Parameters	Field Methods
1. Is the project still functioning as designed?		Fish passage restoration project is within passage guidelines.	
a. Is there still a sufficient jump pool depth for targeted species and life stages?	Residual pool depth at downstream outlet (if culvert outlet is perched or has entry leap)	If there is a jump, pool depth is appropriate for leap height. (Not required for no entry leap.)	Thalweg profile through culvert plus water depths
b. Are leap heights still within jumping ability for targeted species and life stages?	Leap height (residual pool water surface elevation to passage outlet)	Leap height is below critical heights for targeted species and life stage. (Not applicable for no entry leap.)	Thalweg profile through culvert
c. Is stream velocity in critical flow areas still within the swimming ability of the target species and life stages?	Stream velocity in critical area	Stream velocity is equal to or less than swimming ability of target species and life stage.	Stream velocity/discharge measurements
d. Is upstream inlet of the passage area/structure still at grade or below the channel bed?	Bed elevation at inlet and inlet elevation	Culvert inlet matches grade of the natural channel bed.	Thalweg profile through culverts
e. Is the passage area/ structure still at grade?	Slope	Passage structure is at specific designed slope or the slope relative to the natural channel.	Thalweg profile through culvert
f. Can sediment bedload still pass through the restored area?	Slope (top riffle to opening), active channel width, hydraulic capacity.	Passage inlet shows no signs of clogging or deposition.	Thalweg profile through culverts, Cross section surveys
g. Can the structure pass the design flood discharge and meet headwater policies?	Hydraulic capacity	Passage passes 100-yr flows and watershed products.	Cross section surveys
h. Does the passage project show signs of imminent failure?	Structural integrity	Structure shows no signs of collapsing.	Inspection of all culvert structural elements
2. Have channel or bank adjustments impaired the function of the passageway?	Slope, head-cutting, sediment deposition	Channel adjustments have not impaired passage or habitat values.	Thalweg profile through culverts
3. Did the project have adverse effects on upstream or downstream habitat?	Bank erosion, channel incision/head-cutting, debris accumulation or sediment deposition	Passage project has not adversely affected up and downstream habitat.	Thalweg profile through culverts, Cross section surveys
4. Is upstream habitat still suitable for the targeted fish species and life stages?	Habitat types and quality in upstream reaches	Area is still suitable for targeted species and life stages.	Habitat monitoring

9.3 MAINTENANCE

Maintenance activities may result from a regularly scheduled inspection/maintenance program or arise on an emergency basis. Remedial maintenance may be triggered by inspection results. The inspection report should identify and prioritize maintenance needs that are not emergencies, but require attention. Emergency maintenance requires immediate mobilization to repair or prevent damage. Some maintenance actions will require permits, and such requirements should be identified well in advance to accommodate the permitting process. Similarly, access to areas likely to require maintenance should be established at the time of construction.

This page intentionally left blank.

CHAPTER 10 - REFERENCES

Aaserude, R., and Orsborn, J., 1985. "New Concepts in Fish Ladder Design, Volume II of IV; Results of Laboratory and Field Research on New Concepts in Weir and Pool Fishways." Rep. No. DOE/BP-36523-3, Bonneville Power Administration.

Admiraal, D., and Schainost, S., 2004. "Fish Passage for Warm Water Fish Species." World Water and Environmental Resources Congress, 6-27-2004, Salt Lake City, Utah.

Alaska Department of Fish and Game, 2005. "Fish Passage Improvement Program," http://www.sf.adfg.state.ak.us/SARR/fishpassage/fishpass.cfm accessed January 25.

Alaska Department of Fish and Game and Alaska Department of Transportation, 2001. "Memorandum of Agreement Between Alaska Department of Fish and Game and Alaska Department of Transportation and Public Facilities for the Design, Permitting, and Construction of Culverts for Fish Passage," Juneau, Alaska.

Andrews, E.D., 1983. "Entrainment of gravel from naturally sorted riverbed material," Geological Society of America Bulletin, 94, 1184-92.

Arcement, G.K. and V.R. Schneider, 1984. "Guide for Selecting Manning's Roughness Coefficients for Natural Channels and Flood Plains," FHWA Report No., FHWA-TS-84-204, Washington, D.C.

Archfield, Stacey A. and Richard M. Vogel. (2008). "A decision-support system to assess surface-water resources in Massachusetts." proceedings, World Environmental and Water Resources Congress.

Archfield, Stacey A., R. M. Vogel, and Sara L. Brandt. (2007). "Estimation of flow-duration curves at ungaged sites in southern New England." Proceedings, World Environmental and Water Resources Congress. Karen C. Kabbes, editor.

Archfield, Stacey A., 2007. Personal communication (12/4/07)

Ashton, W. S., 1984. "Determination of Seasonal, Frequency and Durational Aspects of Streamflow with Regard to Fish Passage Through Roadway Drainage Structures." Rep. No. AK-RD-85-06, State of Alaska Department of Transportation and Public Facilities Division of Planning and Programming. Fairbanks, AK, University of Alaska, Fairbanks.

Bainbridge, R., 1959. "Speed and Stamina in Three Fish." Journal of Experimental Biology, 37(1), 129-153.

Baker, C. O., and Votapka, F. E., 1990. "Fish Passage Through Culverts." Rep. No. FHWA-FL-90-006, San Dimas, California, Forest Service Technology and Development Center.

Ballinger, C. A., and Drake, P. G., 1995. "Culvert Repair Practices Manual: Volume I." Rep. No. FHWA-RD-94-096, Federal Highway Administration.

Barber, M. E., and Downs, R. C., 1996. "Investigation of Culvert Hydraulics Related to Juvenile Fish Passage." Rep. No. WA-RD 338.1, Washington State Transportation Center (TRAC).

Barnes, Harry H. Jr., 1967. "Roughness Characteristics of Natural Channels." U.S. Geological Survey Water Supply Paper 1849. U.S. Government Printing Office, Washington, D.C.

Bates, K. K., B. Barnard, B. Heiner, P. Klavas, and P. D. Powers, 2003. "Design of Road Culverts for Fish Passage." Washington Department of Fish and Wildlife.

Bates, Kozmo Ken and Rich Kirn, 2007. "Guidelines for the Design of Stream/Road Crossings for Passage of Aquatic Organisms in Vermont," Vermont Fish and Wildlife Department.

Bates, Kozmo Ken and Michael Love, 2009. "Design of Culvert Retrofits for Fish Passage," United States Forest Service, July, Draft.

Bathurst, J.C., 1987, "Critical conditions for bed material movement in steep, boulder-bed streams," in Erosion and Sedimentation in the Pacific Rim. Wallingford: Institute of Hydrology, IAHS Publication, 133, 91-6.

Bathurst, J. C., Graf, W. H., and Cao, H. H., 1987. "Bed Load Discharge Equations for Steep Mountain Rivers." Sediment Transport in Gravel-bed Rivers, C. R. Thorne, J. C. Bathurst, and R. D. Hey, eds., John Wiley & Sons Ltd.

Bathurst, J.C. R.M. Li, and D.B. Simons, 1981. "Resistance Equation for Large-Scale Roughness." Journal of the Hydraulics Division, ASCE, Vol. 107, No. HY12, Proc. Paper 14239, December, pp. 1593-1613.

Behlke, C. E., Kane, D. L., McClean, R. F., and Travis, M. D., 1989. "Field Observations of Arctic Grayling Passage Through Highway Culverts." Transportation Research Record, (1224), 63-66.

Behlke, C. E., Kane, D. L., McClean, R. F., and Travis, M. D., 1991. "Fundamentals of Culvert Design for Passage of Weak-Swimming Fish." Rep. No. FHWA-AK-RD-90-10, U.S. Department of Transportation Federal Highway Administration.

Bell, M. C., 1986. "Fisheries Handbook of Engineering Requirements and Biological Criteria." U.S. Army Corps of Engineers, Fish Passage Development and Evaluation Program, Portland, OR.

Blodgett, J.C., 1986. "Rock Riprap Design for Protection of Stream Channels Near Highway Structures, Volume 1 - Hydraulic Characteristics of Open Channels," USGS Water Resources Investigations Report 86-4127.

Bradley, J.B., D.L. Richards, and C.D. Bahner, 2005. "Debris Control Structures – Evaluation and Countermeasures," Hydraulic Engineering Circular 9 (HEC 9), Third Edition, FHWA-IF-04-016, Federal Highway Administration.

Brown, Scott A. and Eric S. Clyde, 1989. "Design of Riprap Revetment," Hydraulic Engineering Circular Number 11 (HEC 11), FHWA-IP-89-016, Federal Highway Administration.

Browning, M., 1990. "Oregon Culvert Fish Passage Survey." Western Federal Lands Highway Division, Federal Highway Administration, Vancouver, WA.

Brunner, Gary W., 2008. "HEC-RAS River Analysis System User's Manual," U.S. Army Corps of Engineers, March.

Brunner, Gary W. and Matthew J. Fleming, 2009. "HEC-SSP Statistical Software Package," U.S. Army Corps of Engineers, CPD-86, April.

Bunte, K., and Abt, S. R., 2001. "Sampling Surface and Subsurface Particle-Size Distributions in Wadable Gravel- and Cobble-Bed Streams for Analyses in Sediment Transport, Hydraulics, and Streambed Monitoring." Rep. No. RMRS-GTR-74, United States Department of Agriculture Forest Service, Rocky Mountain Research Station.

California Department of Transportation (CALTRANS), 2006. "Design Information Bulletin No. 83-01," CALTRANS Supplement to FHWA Culvert Repair Practices Manual.

California Department of Transportation (CALTRANS). 2007. Fish passage design for road crossings.

Carling, P.A., 1992. "In-stream Hydraulics and Sediment Transport," in "The Rivers Handbook: Hydrological and Ecological Principles," edited by Peter Calow and Geoffrey E. Petts, Blackwell Scientific Publications.

Castellarin, Attilio, Giorgio Camorani, and Armando Brath. (2007). "Predicting annual and long-term flow-duration curves in ungauged basins." Advances in Water Resources, 30(2007), 937-953

Castro, Janine., 2003. "Geomorphic Impacts of Culvert Replacement and Removal: Avoiding Channel Incision." USFWS - Oregon Fish and Wildlife Office, Portland, OR.

Chow, V.T. 1959. "Open Channel Hydraulics," New York, Company, McGraw-Hill Book

Clancy, C. G., 1990. "A Detachable Fishway for Steep Culverts." North American Journal of Fisheries Management, 10(2), 244-246.

Clarkin, K., M.J. Furniss, B. Gubernick, M. Love, K. Moynan, and S.W. Musser, 2003. "National Inventory and Assessment Procedure For Identifying Barriers to Aquatic Organism Passage at Road-Stream Crossings." U.S. Forest Service San Dimas Technology and Development Center, San Dimas, CA.

Clinton, W.J., 1995. "Executive Order 12962: Recreational Fisheries." Federal Register, 60(111).

Coffman, J. S., 2005. "Evaluation of a Predictive Model for Upstream Fish Passage Through Culverts." Master of Science Biology, James Madison University.

Committee on Restoration of Aquatic Ecosystems, 1992. "Restoration of Aquatic Ecosystems: Science, Technology, and Public Policy." National Research Council, Washington, D.C.

Cowan, W. L., 1956. "Estimating Hydraulic Roughness Coefficients," Agricultural Engineering, Volume 37, No. 7.

Dane, B. G., 1978. "A Review & Resolution of Fish Passage Problems at Culvert Sites in British Columbia." Fisheries and Marine Service Technical Report No. 810, Department of Fisheries and Environment.

Doyle, Martin W., Doug Shields, Karin F. Boyd, Peter Skidmore, and DeWitt Dominick. 2007. Channel forming discharge selection in river restoration design. Journal of Hydraulic Engineering 133(7): 831-837, July.

Ellis, J. E., 1974. "The Jumping Ability and Behavior of Green Sunfish (Lepomis cyanellus) at the Outflow of a 1.6-ha Pond." Transactions of the American Fisheries Society, 103(3), 620-623.

Evans, W. A., and Johnston, F. B., 1972. "Fish Migration and Fish Passage: A Practical Guide to Solving Fish Passage Problems." U.S.D.A. Forest Service - Region 5.

Everest, F. H., Sedell, J. R., Armantrout, N. B., Nickerson, T. E., Keller, S. M., Johnson, J. M., Parante, W. D., and Haugen, G. N., 1985. "Salmonids." Management of Wildlife and Fish Habitats in Forests of Western Oregon and Washington - Part 1, Brown E.R., ed., U.S.D.A. Forest Service, 199-230.

Farhig, L., and Merriam, G., 1985. "Habitat Patch Connectivity and Population Survival." Ecology, 66(6), 1762-1768.

Federal Highway Administration, 2004. "National Bridge Inspection Standards." Federal Register, 69(239), 74419-74439.

Fennessey, Neil and Richard M. Vogel. (1990). "Regional flow-duration curves for ungauged sites in Massachusetts." Journal of Water Resources Planning and Management, 116(4), 530 - 549.

Flanagan, S. A., Furniss, M., Ledwith, T. S., Thiesen, S., Love, M., Moore, K., and Ory, J., 1998. "Methods for Inventory and Environmental Risk Assessment of Road Drainage Crossings." Rep. No. 9877 1809P - SDTDC, USDA Forest Service.

Flynn, Kathleen M., William H. Kirby, and Paul R. Hummel, 2006. "User's Manual for Program PeakFQ, Annual Flood-Frequency Analysis Using Bulletin 17B Guidelines," U.S. Department of the Interior, U.S. Geological Survey.

Forest Practices Advisory Committee on Salmon in Watersheds, 2001. "Section A: Fish Passage Restoration."

Forest Service Stream Simulation Working Group (FSSWG), 2008. "Stream Simulation: An Ecological Approach to Road-Stream Crossings, May.

Fuller, W.B. and S.E. Thompson, 1907. "The Laws of Proportioning Concrete," Journal of the Transportation Division, American Society of Civil Engineers, Volume 59, 67-143.

Furniss, M., 2006. "Recognizing and Avoiding Errors in Culvert AOP Practice." Fish Passage Summit Meeting, 2-16-2006, Denver, CO.

Gebhards, S., and Fisher, J., 1972. "Fish Passage and Culvert Installations." Idaho Fish & Game Department.

General Accounting Office, 2001. "Restoring Fish Passage Through Culverts on Forest Service and BLM Lands in Oregon and Washington Could Take Decades." Rep. No. GAO-02-136, General Accounting Office.

Gerking, S. D., 1959. "The Restricted Movement of Fish Populations." Cambridge Biological Society Biological Review, 34, 221-242.

Gessler, Johannes (1965). "The beginning of bed load movement of mixtures investigated as natural armoring in channels," Report Number 69 of the Laboratory of Hydraulic Research and Soil Mechanics of the Swiss Federal Institute of Technology.

Gowan, C., Young, M. K., Fausch, K. D., and Riley, S. C., 1994. "Restricted Movement in Resident Stream Salmonids: A Paradigm Lost?" Canadian Journal of Aquatic Science, 51(11), 2626-2637.

Gregory, S., McEnroe, J., Klingeman, P., and Wyrick, J., 2004. "Fish Passage Through Retrofitted Culverts." Rep. No. FHWA-OR-RD-05-05, Federal Highway Administration.

Groot, C., and Margolis, L., 1991. "Pacific Salmon Life Histories." University of British Columbia Press, Vancouver, B.C.

Gubernick, R., Tongass National Forest, Personal Communication, 1-24-2007.

Harrelson, C. C., Rawlins, C. L., and Potyondy, J. P., 1994. "Stream Channel Reference Sites: An Illustrated Guide to Field Technique." Gen. Tech. Rep. No. RM-245. USDA Forest Service, Rocky Mountain Forest and Range Experiment Station, Fort Collins, CO.

Harris, R. R., 2005. "Monitoring the Effectiveness of Culvert Fish Passage Restoration - Final Report." Center for Forestry, University of California, Berkeley.

Hinch, S. G., and Rand, P. S., 1998. "Swim Speeds and Energy Use of Upriver-Migrating Sockeye Salmon (Oncorhynchus nerka): Role of Local Environment and Fish Characteristics." Canadian Journal of Fisheries and Aquatic Science, 55(8), 1821-1831.

Hotchkiss, Rollin H. and Christopher M. Frei, 2007. "Design for Fish Passage and Road-Stream Crossings: Synthesis Report," Federal Highway Administration.

Hudy, M., 2006. "A Need for Coarse Filters?" CD Rom Proceedings, Fish Passage Summit Meeting, 2-16-2006, Denver, Colorado.

Jackson, S., 2003. "Design and Construction of Aquatic Organism Passage at Road-Stream Crossings: Ecological Considerations in the Design of River and Stream Crossings." 20-29 International Conference of Ecology and Transportation, Lake Placid, New York

Jarrett, R.D., 1984. "Hydraulics of High Gradient Streams," Journal of the Hydraulics Division, ASCE, Volume 110(11).

Johnson, P., and Brown, E., 2000. "Stream Assessment for Multicell Culvert Use." Journal of Hydraulic Engineering, 126(5), 381-386.

Jones, D. R., Kiceniuk, J. W., and Bamford, O. S., 1974. "Evaluation of the Swimming Performance of Several Fish Species from the Mackenzie River." Journal of the Fisheries Research Board of Canada, 31, 1641-1647.

Kahler, T., and Quinn, T., 1998. "Juvenile and Resident Salmonid Movement and Passage through Culverts." Rep. No. WA-RD 457.1.

Kemp, P. S., Gessel, M. H., Sandford, B. P., and Williams, J. G., 2006. "The Behaviour of Pacific Salmonid Smolts During Passage Over two Experimental Weirs Under Light and Dark Conditions." River Research and Applications, 22(4), 429-440.

Kilgore, Roger T. and George K. Cotton, 2005. "Design of Roadside Channels with Flexible Linings," Hydraulic Engineering Circular Number 15 (HEC 15), 3rd Edition, FHWA-NHI-05-114, Federal Highway Administration.

Kilgore, Roger T., and G. K. Young, 1993. "Riprap Incipient Motion and Shields' Parameter." Proceedings of the American Society of Civil Engineers' Hydraulics Division Conference.

Komar, P.D., 1987. "Selective grain entrainment and the empirical evaluation of flow competence," Sedimentology 34, 1165-1176.

Komar, P.D., 1996. "Entrainment of sediments from deposits of mixed grain sizes and densities," in Carling, P.A. and Dawson, M.R. (eds), Advances in Fluvial Dynamics and Stratigraphy: Wiley, Chichester, 127-81.

Komar, P.D. and P.A. Carling, 1991. "Grain sorting in gravel-bed streams and the choice of particle sizes for flow-competence evaluations," Sedimentology, 38, 489-502.

Lagasse, P.F., P.E. Clopper, L.W. Zevenbergen, and J.R. Ruff, 2006. "Riprap Design Criteria, Recommended Specifications, and Quality Control," NCHRP Report 568, Transportation Research Board.

Lagasse, P.F., J.D. Schall, and E.V. Richardson, 2001a. "Stream Stability at Highway Structures," Hydraulic Engineering Circular Number 20 (HEC 20), Third Edition, FHWA-NHI-01-002, Federal Highway Administration.

Lagasse, P.F., L.W. Zevenbergen, J.D. Schall, and P.E. Clopper, 2001b. "Bridge Scour and Stream Instability Countermeasures," Hydraulic Engineering Circular Number 23 (HEC 23), Second Edition, FHWA-NHI-01-003, Federal Highway Administration.

Lane, E.W., 1955. "The Importance of Fluvial Geomorphology in Hydraulic Engineering," ASCE Proceedings, Volume 81, No. 745, pp. 1-17.

Lang, M., Love, M., and Trush, W., 2004. "Improving Stream Crossings for Fish Passage." Rep. No. 50ABNF800082, National Marine Fisheries Service.

Leopold, L. B., Wolman, M. G., and Miller, J. P., 1964. "Fluvial Processes in Geomorphology." Freeman, San Francisco, CA.

Limerinos, J. T., 1970. "Determination of the Manning Coefficient From Measured Bed Roughness in Natural Channels." Geological Survey Water-Supply Paper 1898-B, United States Government Printing Office, Washington, D.C.

Maine Department of Transportation, 2004. "Fish Passage Policy and Design Guide: 2nd Edition." Maine Department of Transportation.

Maryland State Highway Administration, 2005. "Culverts, Guidelines for the Selection and Design of Culvert Installations," Manual for Hydrologic and Hydraulic Design, Office of Bridge Development.

McCuen, Richard H., Peggy A. Johnson, and Robert M. Ragan, 2002. "Highway Hydrology," Hydraulic Design Series Number 2 (HDS 2), FHWA-NHI-02-001, Federal Highway Administration.

McCullah, John and Donald Gray, 2005. "Environmentally Sensitive Channel and Bank Protection Measures," NCHRP Report 544, Transportation Research Board.

Montgomery, D. R., and Buffington, J. M., 1993. "Channel Classification, Prediction of Channel Response, and Assessment of Channel Condition." Rep. No. TFW-SI-110-93-002.

Montgomery, D. R., and Buffington, J. M., 1998. "Channel Classification, Processes, and Response." Ecology and Management in Pacific Northwest Rivers.

Mussetter, R. A., 1989. "Dynamics of Mountain Streams." Doctorate of Philosophy, Colorado State University.

Normann, J. M., Houghtalen, R. J., and Johnston, W. J., 2005. "Hydraulic Design Series No. 5, 2nd Edition, rev.: Hydraulic Design of Highway Culverts." Rep. No. FHWA-NHI-01-020, Federal Highway Administration.

Powers, P.D. and J.F. Orsborn, 1985. "Analysis of Barriers to Upstream Fish Migration," Project No. 82-14, Bonneville Power Administration, Portland Oregon.

Powers, P. and C. Saunders. 1996. Fish passage design flows for ungaged catchments in Washington. Washington Department of Fish and Wildlife.

Powers, P., Bates K., Gowen, B., and Whitney, R., 1997. "Culvert Hydraulics Related to Upstream Juvenile Salmon Passage." Washington State Department of Transportation, Olympia, WA.

Rajaratnam, N., Katopodis, C., and Lodewyk, S., 1991. "Hydraulics of Culvert Fishways IV: Spoiler Baffle Culvert Fishways." Canadian Journal of Civil Engineering, 18(1), 76-82.

Rand, P. S., and Hinch, S. G., 1998. "Swim Speeds and Energy Use of Upriver-Migration Sockeye Salmon (Oncorhynchus nerka): Simulating Metabolic Power and Assessing Risk of Energy Depletion." Canadian Journal of Fisheries and Aquatic Science, 55(8), 1832-1841.

Richardson, E.V., D.B. Simons, and P.F. Lagasse, 2001. "River Engineering for Highway Encroachments," Hydraulic Design Series Number 6 (HDS 6), FHWA-NHI-01-004, Federal Highway Administration.

River and Stream Continuity Partnership, 2004. "Massachusetts River and Stream Crossing Standards: Technical Guidelines."

Robison, E.G., A. Mirati, and M. Allen, 1999. "Oregon Road/Stream Crossing Restoration Guide: Spring 1999."

Rosgen, D. L., 1994. "A Classification of Natural Rivers." Catena, 22(3), 169 □ 199.

Rosgen, D. L., 1996. "Applied River Morphology." Wildland Hydrology Press, Pagosa Springs, CO.

Rowland, E.R., Hotchkiss, R.H., and Barber, M.E., 2003a. "Modeling Hydrology for Fish Passage." Rep. No. WA-RD 545.1, Washington State Department of Transportation, Olympia, WA.

Rowland, E.R., Hotchkiss, R.H., and Barber, M.E., 2003b. "Predicting fish passage design flows at ungaged streams in Eastern Washington," Journal of Hydrology, 273: 177-187.

Schaefer, J. F., Marsh-Matthews, E., Spooner, D. E., Gido, K. B., and Matthews, W. J., 2003. "Effects of Barriers and Thermal Refugia on Local Movement of the Threatened Leopard Darter, Percina pantherina." Environmental Biology of Fishes, 66(4), 391-400.

Schrag, A. M., 2003. "Highways and Wildlife: Review of Mitigation Projects throughout Europe, Canada and the United States." Masters of Science, California State Polytechnic University, Pomona, CA.

Schumm, S.A., M.D. Harvey, and C.C. Watson, 1984. "Incised Channels: Morphology, Dynamics and Control," Water Resources Publications, Littleton, Colorado, USA.

Scott, W. B., and Crossman, E. J., 1973. "Freshwater Fishes of Canada." Fisheries Research Board of Canada, Ottawa.

Smith, D. L., and Brannon, E. L., 2006. "Use of Average and Fluctuating Velocity Components for Estimation of Volitional Rainbow Trout Density." Transactions of the American Fisheries Society, 135, 431-441.

Spellerberg, I. F., 1998. "Ecological Effects of Roads and Traffic: A Literature Review." Global Ecology and Biogeographical Letters, 7(5), 317-333.

Strickler, Alfred, 1923. "Some Contributions to the Problem of the Velocity Formula and Roughness Factors for Rivers, Canals, and Closed Conduits," Bern, Switzerland, Mitt. Eidgeno Assischen Amtes Wasserwirtschaft.

Stuart, T. A., 1962. "The Leaping Behaviour of Salmon and Trout at Falls and Obstructions." Department of Agriculture and Fisheries for Scotland.

Studley, Seth E. 2000. Estimated flow-duration curves for ungaged sites in the Cimarron and lower Arkansas River basins in Kansas. U.S. Geological Survey Water-Resources Investigations Report 00-04113.

Taylor, R. N., and Love, M., 2003. "Part IX Fish Passage Evaluation at Stream Crossings." California Department of Fish and Game.

Thurman, David R. and Alex R. Horner-Devine, 2007. "Hydrodynamic Regimes and Structures in Sloped Weir Baffled Culverts and Their Influence on Juvenile Salmon Passage," Washington State Department of Transportation, WA-RD 687.1, November.

Toepfer, C. S., Fisher, W. L., and Haubelt, J. A., 1999. "Swimming Performance of the Threatened Leopard Darter in Relation to Road Culverts." Transactions of the American Fisheries Society, 128(1), 155-161.

Trombulak, S. C., and Frissell, C. A., 2000. "Review of Ecological Effects of Roads on Terrestrial and Aquatic Communities." Conservation Biology, 14(1), 18-30.

United States Congress, 1899. "Rivers and Harbors Appropriation Act." 33 U.S.C. 403; Chapter 425, March 3, 1899; 30 Stat. 1151. Washington: GPO.

United States Congress, 1934. "Fish and Wildlife Coordination Act." 16 U.S.C. 661-667e; the Act of March 10, 1934. Washington, GPO.

United States Congress, 1977. "Clean Water Act" CWA; 33 U.S.C. 1251-1376. Washington: GPO.

United States Congress. 1968. "Wild and Scenic Rivers Act." 16 U.S.C. 1271-1287 -- Public Law 90-542, approved October 2, 1968, (82 Stat. 906). Washington: GPO.

United States Congress, 1969. "National Environmental Policy Act." Public Law 91-190, 42 U.S.C. 4321-4347, January 1, 1970. Washington: GPO.

United States Congress, 1973. "Endangered Species Act." 16 U.S.C. 1531-1544, 87 Stat. 884, as amended. Washington: GPO.

United States Congress, 1996. "Magnuson-Stevens Fishery Conservation and Management Act." PL 94-265 (16 U.S.C.), as amended. Washington: GPO.

U.S. Department of Agriculture, 1987. "Stability of grassed-lined open channels," Agricultural Research Service, Agricultural Handbook Number 667.

United States Forest Service, 2006a. "Course Notes." Designing for Aquatic Organism Passage at Road-Stream Crossings Short Course, 2-6-2006, Pacific City, OR

United States Forest Service, 2006b. "FishXing Case Histories," http://www.stream.fs.fed.us/fishxing/case.html accessed October 5.

Venner Consulting and Parsons Brinkerhoff, 2004. "Environmental Stewardship Practices, Procedures, and Policies for Highway Construction and Maintenance." Final Report for NCHRP Project 25-25, Task 4, National Cooperative Highways Research Program Transportation Research Board.

Wang, S.Y. and H.W. Shen, 1985. "Incipient Sediment Motion and Riprap Design." Journal of Hydraulics Division, ASCE, Vol. 111, No. 3, March, pp. 52-538.

Washington Department of Fish and Wildlife, 2000. "Fish Passage Barrier and Surface Water Diversion Screening Assessment and Prioritization Manual."

Weaver, C. R., Thompson, C. S., and Slatick, E., 1976. "Fish Passage Research at the Fisheries-Engineering Research Laboratory May 1965 to September 1970." Rep. No. 32, North Pacific Division, U.S. Army Corps of Engineers.

Webb, P. W., 1975. "Hydrodynamics and Energetics of Fish Propulsion Bulletin 190." Ottawa, Ontario.

Welton, J. S., Beaumont, W. R. C., and Clarke, R. T., 2002. "The Efficacy of Air, Sound and Acoustic Bubble Screens in Deflecting Atlantic Salmon, Salmo salar L., Smolts in the River Frome, UK." Fisheries Management and Ecology, 9(1), 11-18.

White, D., 1997. "Hydraulic Performance of Countersunk Culverts in Oregon." Master of Science, Oregon State University.

Wyant, David C., 2002. "Assessment and Rehabilitation of Existing Culverts: A Synthesis of Highway Practice," NCHRP Synthesis 303, Transportation Research Board, Washington, D.C.

APPENDIX A- METRIC SYSTEM, CONVERSION FACTORS, AND WATER PROPERTIES

The following information is summarized from the Federal Highway Administration, National Highway Institute (NHI) Course No. 12301, "Metric (SI) Training for Highway Agencies." For additional information, refer to the Participant Notebook for NHI Course No. 12301.

In SI there are seven base units, many derived units and two supplemental units (Table A.1). Base units uniquely describe a property requiring measurement. One of the most common units in civil engineering is length, with a base unit of meters in SI. Decimal multiples of meters include the kilometer (1000 m), the centimeter (1 m/100) and the millimeter (1 m/1000). The second base unit relevant to highway applications is the kilogram, a measure of mass that is the inertia of an object. There is a subtle difference between mass and weight. In SI, mass is a base unit, while weight is a derived quantity related to mass and the acceleration of gravity, sometimes referred to as the force of gravity. In SI the unit of mass is the kilogram and the unit of weight/force is the Newton. Table A.2 illustrates the relationship of mass and weight. The unit of time is the same in SI as in the Customary (English) system (seconds). The measurement of temperature is Centigrade. The following equation converts Fahrenheit temperatures to Centigrade, °C = 5/9 (°F - 32).

Derived units are formed by combining base units to express other characteristics. Common derived units in highway drainage engineering include area, volume, velocity, and density. Some derived units have special names (Table A.3).

Table A.4 provides the standard SI prefixes and their definitions. Table A.5 provides useful conversion factors from Customary to SI units. The symbols used in this table for metric (SI) units, including the use of upper and lower case (e.g., kilometer is "km" and a Newton is "N") are the standards that should be followed.

Table A.6 provides physical properties of water at atmospheric pressure in SI units, A.7 in customary units. Table A.8 gives the sediment grade scale. Table A.9 and A.10 provide common equivalent hydraulic units.

Table A.1.   Overview of SI.

		Units	Symbol
Base units	Length	meter	m
	Mass	kilogram	kg
	Time	second	s
	temperature*	kelvin	K
	electrical current	ampere	A
	luminous intensity	candela	cd
	amount of material	mole	mol
Derived units			
Supplementary units	angles in the plane	radian	rad
	solid angles	steradian	sr

* Use degrees Celsius (°C), which has a more common usage than kelvin.

Table A.2. Relationship of Mass and Weight.

	Mass	Weight or Force of Gravity	Force
Customary	slug	pound	pound
	pound-mass	pound-force	pound-force
Metric	kilogram	newton	newton

Table A.3. Derived Units with Special Names.

Quantity	Name	Symbol	Expression
Frequency	hertz	Hz	s^{-1}
Force	newton	N	$kg \cdot m/s^2$
Pressure, stress	pascal	Pa	N/m^2
Energy, work, quantity of heat	joule	J	$N \cdot m$
Power, radiant flux	watt	W	J/s
Electric charge, quantity	coulomb	C	$A \cdot s$
Electric potential	volt	V	W/A
Capacitance	farad	F	C/V
Electric resistance	ohm	Ω	V/A
Electric conductance	siemens	S	A/V
Magnetic flux	weber	Wb	$V \cdot s$
Magnetic flux density	tesla	T	Wb/m^2
Inductance	henry	H	Wb/A
Luminous flux	lumen	lm	$cd \cdot sr$
Illuminance	lux	lx	lm/m^2

Table A.4. Prefixes.

Submultiples			Multiples		
Deci	10^{-1}	d	deka	10^1	da
Centi	10^{-2}	c	hector	10^2	h
Milli	10^{-3}	m	kilo	10^3	k
Micro	10^{-6}	μ	mega	10^6	M
Nano	10^{-9}	n	giga	10^9	G
Pica	10^{-12}	p	tera	10^{12}	T
Femto	10^{-15}	f	peta	10^{15}	P
Atto	10^{-18}	a	exa	10^{18}	E
Zepto	10^{-21}	z	zeta	10^{21}	Z
Yocto	10^{-24}	y	yotto	10^{24}	Y

Table A.5. Useful Conversion Factors.

Quantity	From English Units	To Metric Units	Multiplied by*
Length	mile	Km	1.609
	yard	M	0.9144
	foot	M	0.3048
	inch	Mm	25.4
Area	square mile	km²	2.590
	acre	m²	4047
	acre	hectare	0.4047
	square yard	m²	0.8361
	square foot	m²	0.092 90
	square inch	mm²	645.2
Volume	acre foot	m³	1 233
	cubic yard	m³	0.7646
	cubic foot	m³	0.028 32
	cubic foot	L (1000 cm³)	28.32
	100 board feet	m³	0.2360
	gallon	L (1000 cm³)	3.785
	cubic inch	cm³	16.39
Mass	lb	Kg	0.4536
	kip (1000 lb)	metric ton (1000 kg)	0.4536
Mass/unit length	plf	kg/m	1.488
Mass/unit area	psf	kg/m²	4.882
Mass density	pcf	kg/m³	16.02
Force	lb	N	4.448
	kip	kN	4.448
Force/unit length	plf	N/m	14.59
	klf	kN/m	14.59
Pressure, stress, modulus of elasticity	psf	Pa	47.88
	ksf	kPa	47.88
	psi	kPa	6.895
	ksi	MPa	6.895
Bending moment, torque, moment of force	ft-lb	N А m	1.356
	ft-kip	kN А m	1.356
Moment of mass	lb · ft	kg · m	0.1383
Moment of inertia	lb · ft²	kg · m²	0.042 14
Second moment of area	In⁴	mm⁴	416 200
Section modulus	In³	mm³	16 390
Power	ton (refrig)	kW	3.517
	Btu/s	kW	1.054
	hp (electric)	W	745.7
	Btu/h	W	0.2931
Volume rate of flow	ft³/s	m³/s	0.028 32
	cfm	m³/s	0.000 471 9
	cfm	L/s	0.4719
	mgd	m³/s	0.0438
Velocity, speed	ft/s	m/s	0.3048
Acceleration	f/s²	m/s²	0.3408
Momentum	lb · ft/sec	kg · m/s	0.1383
Angular momentum	lb · ft²/s	kg · m²/s	0.042 14
Plane angle	Degree	Rad	0.017 45
		mrad	17.45

* 4 significant figures; underline denotes exact conversion

Table A.6. Physical Properties of Water at Atmospheric Pressure (SI Units).

Temperature		Density	Specific Weight	Dynamic Viscosity	Kinematic Viscosity	Vapor Pressure	Surface Tension[1]	Bulk Modulus
Centigrade	Fahrenheit	kg/m^3	N/m^3	$N \cdot s/m^2$	m^2/s	N/m^2 abs.	N/m	GN/m^2
0°	32°	1,000	9,810	1.79×10^{-3}	1.79×10^{-6}	611	0.0756	1.99
5°	41°	1,000	9,810	1.51×10^{-3}	1.51×10^{-6}	872	0.0749	2.05
10°	50°	1,000	9,810	1.31×10^{-3}	1.31×10^{-6}	1,230	0.0742	2.11
15°	59°	999	9,800	1.14×10^{-3}	1.14×10^{-6}	1,700	0.0735	2.16
20°	68°	998	9,790	1.00×10^{-3}	1.00×10^{-6}	2,340	0.0728	2.20
25°	77°	997	9,781	8.91×10^{-4}	8.94×10^{-7}	3,170	0.0720	2.23
30°	86°	996	9,771	7.97×10^{-4}	8.00×10^{-7}	4,250	0.0712	2.25
35°	95°	994	9,751	7.20×10^{-4}	7.24×10^{-7}	5,630	0.0704	2.27
40°	104°	992	9,732	6.53×10^{-4}	6.58×10^{-7}	7,380	0.0696	2.28
50°	122°	988	9,693	5.47×10^{-4}	5.53×10^{-7}	12,300	0.0679	2.29
60°	140°	983	9,643	4.66×10^{-4}	4.74×10^{-7}	20,000	0.0662	2.28
70°	158°	978	9,594	4.04×10^{-4}	4.13×10^{-7}	31,200	0.0644	2.25
80°	176°	972	9,535	3.54×10^{-4}	3.64×10^{-7}	47,400	0.0626	2.20
90°	194°	965	9,467	3.15×10^{-4}	3.26×10^{-7}	70,100	0.0607	2.14
100°	212°	958	9,398	2.82×10^{-4}	2.94×10^{-7}	101,300	0.0589	2.07

[1]Surface tension of water in contact with air

Table A.7. Physical Properties of Water at Atmospheric Pressure (Customary Units).

Temperature		Density	Specific Weight	Dynamic Viscosity	Kinematic Viscosity	Vapor Pressure	Surface Tension[1]	Bulk Modulus
Fahrenheit	Centigrade	$Slug/ft^3$	lb/ft^3	$lb\text{-}sec/ft^2$	ft^2/sec	lb/in^2	lb/ft	lb/in^2
32°	0°	1.940	62.416	0.374×10^{-4}	1.93×10^{-5}	0.09	0.00518	287,000
40°	4.4°	1.940	62.423	0.323×10^{-4}	1.67×10^{-5}	0.12	0.00514	296,000
50°	10.0°	1.940	62.408	0.273×10^{-4}	1.41×10^{-5}	0.18	0.00508	305,000
60°	15.6°	1.939	62.366	0.235×10^{-4}	1.21×10^{-5}	0.26	0.00504	313,000
70°	21.1°	1.936	62.300	0.205×10^{-4}	1.06×10^{-5}	0.36	0.00497	319,000
80°	26.7°	1.934	62.217	0.180×10^{-4}	0.929×10^{-5}	0.51	0.00492	325,000
90°	32.2°	1.931	62.118	0.160×10^{-4}	0.828×10^{-5}	0.70	0.00486	329,000
100°	37.8°	1.927	61.998	0.143×10^{-4}	0.741×10^{-5}	0.95	0.00479	331,000
120°	48.9°	1.918	61.719	0.117×10^{-4}	0.610×10^{-5}	1.69	0.00466	332,000
140°	60°	1.908	61.386	0.0979×10^{-4}	0.513×10^{-5}	2.89	0.00454	330,000
160°	71.1°	1.896	61.006	0.0835×10^{-4}	0.440×10^{-5}	4.74	0.00441	326,000
180°	82.2°	1.883	60.586	0.0726×10^{-4}	0.385×10^{-5}	7.51	0.00427	318,000
200°	93.3°	1.869	60.135	0.0637×10^{-4}	0.341×10^{-5}	11.52	0.00413	308,000
212°	100°	1.847	59.843	0.0593×10^{-4}	0.319×10^{-5}	14.70	0.00404	300,000

[1] Surface tension of water in contact with air

Table A.8. Sediment Particles Grade Scale.

Size			Approximate Sieve Mesh Opening per Inch		Class
Millimeters	Microns	Inches	Tyler	U.S. Standard	
4000-2000	-	160-80	-	-	Very large boulders
2000-1000	-	80-40	-	-	Large boulders
1000-500	-	40-20	-	-	Medium boulders
500-250	-	20-10	-	-	Small boulders
250-130	-	10-5	-	-	Large cobbles
130-64	-	5-2.5	-	-	Small cobbles
64-32	-	2.5-1.3	-	-	Very coarse gravel
32-16	-	1.3-0.6	-	-	Coarse gravel
16-8	-	0.6-0.3	2 ½	-	Medium gravel
8-4	-	0.3-0.16	5	5	Fine gravel
4-2	-	0.16-0.08	9	10	Very fine gravel
2-1	2000-1000	-	16	18	Very coarse sand
1-1/2	1000-500	-	32	35	Coarse sand
1/2-1/4	500-250	-	60	60	Medium sand
1/4-1/8	250-125	-	115	120	Fine sand
1/8-1/16	125-62	-	250	230	Very fine sand
1/16-1/32	62-31	-	-	-	Coarse silt
1/32-1/64	31-16	-	-	-	Medium silt
1/64-1/128	16-8	-	-	-	Fine silt
1/128-1/256	8-4	-	-	-	Very fine silt
1/256-1/512	4-2	-	-	-	Coarse clay
1/512-1/1024	2-1	-	-	-	Medium clay
1/1024-1/2048	1-0.5	-	-	-	Fine clay
1/2048-1/4096	0.5-0.24	-	-	-	Very fine clay

Table A.9. Common Equivalent Hydraulic Units: Volume.

Unit	Volume							
	Equivalent							
	cubic inch	liter	U.S. gallon	cubic foot	cubic yard	cubic meter	acre-foot	sec-foot-day
Liter	61.02	1	0.264 2	0.035 31	0.001 308	0.001	810.6 E - 9	408.7 E - 9
U.S. gallon	231.0	3.785	1	0.1337	0.004 951	0.003 785	3.068 E - 6	1.547 E - 6
cubic foot	1728	28.32	7.481	1	0.037 04	0.028 32	22.96 E - 6	11.57 E - 6
cubic yard	46,660	764.6	202.0	27	1	0.746 6	619.8 E - 6	312.5 E - 6
cubic meter	61,020	1000	264.2	35.31	1.308	1	810.6 E - 6	408.7 E - 6
acre-foot	75.27 E + 6	1,233,000	325,900	43 560	1.613	1 233	1	0.504 2
sec-foot-day	149.3 E + 6	2,447,000	646,400	86 400	3 200	2 447	1.983	1

Table A.10. Common Equivalent Hydraulic Units: Rates.

Unit	Discharge (Flow Rate, Volume/Time)					
	Equivalent					
	gallon/min	liter/sec	acre-foot/day	foot^3/sec	million gal/day	meter^3/sec
gallon/minute	1	0.063 09	0.004 419	0.002 228	0.001 440	63.09 E - 6
liter/sec	15.85	1	0.070 05	0.035 31	0.022 82	0.001
acre-foot/day	226.3	14.28	1	0.504 2	0.325 9	0.014 28
foot^3/sec	448.8	28.32	1.983	1	0.646 3	0.028 32
million gal/day	694.4	43.81	3.068	1.547	1	0.043 82
meter^3/sec	15,850	1000	70.04	35.31	22.83	1

This page intentionally left blank.

APPENDIX B- LEGISLATION AND REGULATION

Several statutes, regulations and Executive Orders may be relevant for the selection, design, installation, operation, and maintenance of culverts, especially those in waters that support fish. Almost all of the relevant statutes delegate jurisdiction by statute or expertise to one or more regulatory or coordinating agencies.

These statutes and Executive Orders represent societal values and, in most cases, identify obligations of federal agencies that are as important to the public as is a safe and reliable road network. It is a fundamental engineering challenge to collaborate with other disciplines and agencies to identify one or more culvert solutions that optimize as many of those societal values as possible. The information in this section is provided to encourage active and informed interdisciplinary and multiple agency discussions, which will enhance the permitting process, improving cost, time, safety and ecosystem efficiencies.

Environmental regulatory agencies have greatly streamlined and simplified the permit application processes for installing, replacing or extending a culvert at a roadway-stream crossing, but there are still many occasions where the process does not go smoothly, or may be complex and frustrating. Regardless, a key to long-term success is ongoing good faith efforts to help all agencies and stakeholders attain their goals. Striving to meet transportation and environmental goals when roads cross streams requires routine use of common sense, and participation of interdisciplinary and multiple agency teams to support hydraulic, design, safety and structural engineering. This section provides a brief description of some of the most frequently encountered federal environmental statutes.

B.1 FEDERAL STATUTES AND AN EXECUTIVE ORDER

B.1.1 Clean Water Act (CWA) 1977

The Clean Water Act, as it is commonly known, is a 1977 amendment to the Federal Water Pollution Control Act of 1972, which set the basic structure for regulating discharges of pollutants to waters of the United States. The Clean Water Act is intended to restore and maintain the physical, chemical and biological integrity of waters of the United States. Pollutants can include low levels of dissolved oxygen, temperature, sediment and even color. This law is the source of each States' (and some Tribal) water quality standards, which always include an anti-degradation clause: discharge of pollutants cannot degrade the waterway's designated uses. If aquatic life is a designated use, culvert installation, operation and maintenance should not cause physical, chemical or biological degradation or otherwise alter fish species composition and demographics, and habitat. The discharge should not impede fish movements, the movements of prey and forage, or symbiotic and commensal species.

In addition, all states support a list of non-attainment waters as required by CWA 303(d). The 303(d) list is generally linked to total daily maximum load (TMDL) limitations.

Three sections of the CWA are relevant to culvert installation across the country: sections 401 (water quality certification), 402 (National Pollution Discharge Elimination System permits), and 404 (dredge and fill, also called "wetlands" permits). In rare circumstances, Section 403 (ocean discharge permits) may be required. Permits or certification notices issued under sections 401 and 402 may be indistinguishable in practice. They address the project's compliance with State water quality standards; most States, and many tribes, have assumed responsibility for these programs from the Environmental Protection Agency (EPA). Permits issued under Section 404 generally address the placement of fill material, including pipes and the pipe-soil matrix, into designated water bodies. The U.S. Army Corps of Engineers is the primary jurisdictional agency, but the EPA has joint oversight of the program, and Section 404 permits, while most commonly associated with wetlands in the public mind, cover fill in streams, lakes and more.

The regulatory agencies at state and federal levels have established simplified permit processes for routine activities that do not degrade the environment. These may be nationwide, regional or state wide in scope.

B.1.2 Endangered Species Act (ESA) of 1973

The ESA obligates all federal agencies to seek to conserve, or recover, federally listed species, and to use all their authorities and programs, including grants, loans, permit issuance and technical assistance, to do the same (United States Congress, 1973). The law's purpose is to provide a means to conserve the ecosystems on which federally listed species depend, to conserve or recover those listed species, and to meet the Nation's obligations under treaties and conventions. The law and implementing regulations and guidance dictate the process for listing species as threatened or endangered and recognize that federally listed threatened and endangered species are jurisdictionally distinct from State listed species. International and other treaties and conventions may be relevant where certain transboundary fish restoration or invasive species control issues are in effect.

For current purposes, federally listed species fall into two categories. Endangered identifies those species which are in imminent risk of extinction. Threatened identifies the next highest risk category, species or populations facing imminent risk of extirpation. Species that fall under either of two additional categories, proposed and candidate, are not technically considered federally listed. Nevertheless, species that fall within these latter categories generally warrant special administrative procedures or protective measures. The ESA's protections are limited to plants and animals.

Two agencies, the Fish and Wildlife Service and the National Marine Fisheries Service (or National Oceanographic and Atmospheric Administration – Fisheries), collectively called the Services, have jurisdiction by law and expertise. The rationale used to allocate each species and life history stage to a Service is not always clear.

The law also requires jurisdictional agencies to designate critical habitat for listed species. The rulemaking material must include a description of the constituent elements, including structures, processes and ecosystem attributes that must be protected or restored for the habitat to support recovery. This can include geomorphic and hydrologic processes. Federal projects that adversely constrain or alter those constituent elements are said to adversely modify the designated critical habitat.

Federal projects that may affect a listed animal must undergo a cooperative consultation under Section 7 of the ESA with the Fish and Wildlife Service or National Marine Fisheries Service to avoid violating Section 9 prohibitions. This consultation is intended to mitigate the adverse effects of the action on listed species to the extent practicable. Often, agencies that routinely conduct activities that may affect listed species develop a set of best management plans, which preclude the need for formal consultation.

For non-federal entities carrying out an action that may take a listed species, the ESA includes alternate means of working cooperatively with the Services to minimize take and still implement the project without violating Section 9's prohibitions. Take is broadly defined to include harassing, killing, wounding or otherwise interfering with individuals, or disturbing habitat used for feeding, breeding, sheltering and, in the case of fish, spawning and rearing.

Decades of experience suggests that collaboration and ongoing discussions between agencies and disciplines offer the highest level of certainty that consultation for projects that may affect listed fish species will be completed in a timely and effective manner. Because the consultation process is both substantive and procedural, agencies like the Federal Highway Administration recommend using a collaborative, interdisciplinary problem-solving approach to consultation.

B.1.3 Fish and Wildlife Coordination Act (FWCA) 1934

The FWCA recognizes the importance of wildlife resources to the nation (United States Congress, 1934). It requires federal agencies undertaking water resource projects to give equal consideration and coordination to fish and wildlife resource conservation. Undertakings are generally accepted to include funding, permitting and more. The law originally targeted game and fur bearing animals, and commercially and recreationally valuable fish and shellfish (reflecting the traditional concern for "fur, fins and feathers"). Because of emerging scientific knowledge and well-established practice, consideration is now given to ecosystem patterns, processes and the species therein.

The law is one of the vehicles that Fish and Wildlife Service, National Marine Fisheries Service, Department of Agriculture bureaus and State fish and game agencies use to provide cooperative assistance and reports on environmental effects of proposals to Federal action agencies. It authorizes the Fish and Wildlife Service to conduct investigations, including comment letters, to protect environmental resources, and allows Federal agencies to fund preparation of those reports.

B.1.4 National Environmental Policy Act (NEPA) 1969

The NEPA encourages productive and enjoyable harmony between man and the environment as national policy (United States Congress, 1969). It is one of only a few statutes that include the word "ecosystem," and the authors assert that it was introduced and passed to balance the effects of the Full Employment Act of 1948, which obligates Federal agencies to promote economic growth in all of their activities.

More important from the perspective of fish passage and culverts, the NEPA also established the requirement that Federal decisions be informed about the environmental consequences of those actions. These consequences encompass what is described as the human and natural environments. Coverage can be comprehensive when required; the implementing regulations, however, encourage a common sense approach.

The NEPA analysis and documentation is differentiated by four categories, those that are: (1) statutorily excluded, which could include certain disaster response activities, (2) categorically excluded, which include activities that the evidence suggests individually and cumulatively have no significant, lasting effect on the environment, (3) Environmental Assessment and Finding of No Significant Impact, which is usually a cursory or preliminary evaluation of potential effects, with the obvious conclusion, and (4) Environmental Impact Statement and Record of Decision, which are activities that require more substantive analysis because they are likely or known to have significant environmental consequences, or high degrees of uncertainty and controversy.

The NEPA applies to Federal agencies that directly or indirectly implement projects, establish rules or enforce laws. The NEPA analysis is, for example, conducted by the U.S. Army Corps of Engineers when that agency issues a Clean Water Act section 404 permit authorizing installation of a culvert. In many States, the regulatory and decision-making agencies have developed streamlined processes that allow them to join together and conduct a single NEPA analysis for the various agencies that must make decisions. Such streamlined processes provide for "one-stop shopping" to culvert placement proponents, regulatory agencies, environmental agencies and other stakeholders.

The NEPA, in practice, is an important driver for interdisciplinary approaches, public involvement and similar initiatives.

B.1.5 Rivers and Harbors Appropriations Act of 1899

The Rivers and Harbors Act is concerned with navigation in the nations' waters, and the regulation of interstate commerce related to that navigation (United States Congress, 1899). The law has two "permit" sections of interest when considering roadway-stream crossings.

The U.S. Coast Guard, a bureau of the Department of Homeland Security, has jurisdiction over Section 9. This section requires a permit or authorization for construction of bridges, dams, dikes or causeways over or in navigable waterways. An exception in the process exists for navigable waterways that are entirely within one State's boundaries.

The U.S. Army Corps of Engineers has jurisdiction for permits issued under Section 10. Section 10 covers the building of any wharfs, piers, jetties and other structures, and excavation or fill within navigable waters. In practice, Section 10 permits are considered part and parcel of the Clean Water Act section 404 permit process; the popular reference is to a "Section 10/404 permit."

B.1.6 Sustainable Fisheries Act 1996

The Magnuson-Stevens Fishery Conservation and Management Act, also known as the Sustainable Fisheries Act, primarily directs States to work together through various commissions and councils to manage marine and Great Lakes commercial fisheries (United States Congress, 1996). Of interest for the present purposes is the requirement that those multiple state fishery management councils develop fishery management plans, using an ecosystem and ecological approach. The plans must identify Essential Fish Habitat (EFH) for all life stages of the target species and associated species and processes. Coastal waters that are designated EFH under this statute may pose unique challenges when designing, installing, operating and maintaining culverts. The EFH is designated by councils or commissions comprised of state marine fish agencies or their equivalent; EFH within a State are protected by the relevant state agency and the National Marine Fisheries Service, also known as National Oceanographic and Atmospheric Administration – Fisheries. Those agencies review activities authorized or funded by Federal agencies, and coordinate to ensure that the functional integrity of the EFH is not degraded.

B.1.7 Wild and Scenic Rivers Act 1968

The Wild and Scenic Rivers Act establishes a National Wild and Scenic Rivers System (United States Congress, 1968). This law is of interest here only in that special considerations apply when considering culvert installation or any road feature in or adjacent to a designated Wild and Scenic River.

B.1.8 Executive Order on Recreation Fisheries 1995

The Executive Order on Recreational Fisheries (EO), number 12962, directs federal agencies to support recreational fishing (Clinton, 1995). Collaborative efforts are encouraged. Such efforts can include aquatic resource habitat conservation and restoration, implementation of programs in a manner that supports recreational fisheries, and more. The EO can be used to support federal agency involvement in partnerships that address fish passage through culverts.

B.1.9 Executive Order on Floodplain Management 1977

The Executive Order on Floodplain Management, number 11988, requires federal agencies, in part, to restore and preserved the natural and beneficial floodplain values that are adversely impacted by highway agency actions. The requirements are implemented by FHWA in 23 CFR 650, Subpart A.

B.2 STATE AND LOCAL REGULATIONS

In addition to federal regulations, there may also be a number of regional, local or state regulations that apply to the design and installation of roadway-stream crossing structures. For example, the NOAA National Marine Fisheries Service (NMFS) has fish passage policies in place for several of their regions. Such regulations may dictate construction timing, allowable sediment levels, fish passage requirements, or preferred culvert design techniques. It is important to consult with local authorities before beginning any project.

This page intentionally left blank.

APPENDIX C- MANNING'S ROUGHNESS

C.1 CULVERT MATERIAL

Estimation of culvert material n values is well established and needs no further assessment here. Reference is made to Normann, et al. (2005).

C.2 BED AND BANK

A variety of techniques have been proposed for estimating Manning's n values for noncohesive bed and bank materials (no vegetation) including Bathurst et al. (1981), Blodgett (1986), Limerinos (1970), Jarrett (1984), Mussetter (1989), and Strickler (1923). Barnes (1967) photo-documents numerous rivers and streams and reports measured discharges and estimated roughness values. Some of the estimates are for channel bed and non-vegetated banks, while others apply to channels that include vegetated banks and or floodplains. Several FHWA publications also address the topic including HDS 6 (Richardson, et al., 2001), HEC 11 (Brown and Clyde, 1989), and HEC 15 (Kilgore and Cotton, 2005).

Limerinos (1970) proposed several alternative equations based on the D_{50} and D_{84} particle size and measurements of the shortest and intermediate particle axes. The most commonly cited version is for the D_{84} particle size measured on the intermediate axis.

$$n = \frac{\alpha R_h^{1/6}}{1.16 + 2 \log\left(\dfrac{R_h}{D_{84}}\right)} \tag{C.1}$$

where,

n	=	Manning's roughness coefficient, dimensionless
R_h	=	hydraulic radius, ft (m)
D_{84}	=	84th percentile grain size, ft (m)
α	=	0.0926 (CU), 0.1129 (SI)

The Limerinos equation is applicable for $0.9 \le R_h/D_{84} \le 68.5$.

Jarrett (1984) proposed a relationship that did not require specification of bed material size observing that energy slope and particle size are correlated.

$$n = \alpha S_f^{0.38} R_h^{-0.16} \tag{C.2}$$

where,

n	=	Manning's roughness coefficient, dimensionless
R_h	=	hydraulic radius, ft (m)
S_f	=	friction (energy) slope, ft/ft (m/m)
α	=	0.39 (CU), 0.32 (SI)

The Jarrett equation is applicable for $0.002 \le S_f \le 0.04$; 0.5 ft (0.15 m) $\le R_h \le$ 7 ft (2.1 m); and $0.4 \le R_h/D_{84} \le 11$.

Blodgett (1986) proposed a relationship for Manning's roughness coefficient that used the form of the Limerinos equation but is a function of average flow depth and D_{50}, as follows:

$$n = \frac{\alpha \, y_a^{\frac{1}{6}}}{2.25 + 5.23 \log\left(\dfrac{y_a}{D_{50}}\right)} \tag{C.3}$$

where,

n = Manning's roughness coefficient, dimensionless
y_a = average flow depth in the channel, ft (m)
D_{50} = median riprap/gravel size, ft (m)
α = 0.262 (CU), 0.319 (SI)

The Blodgett equation is applicable for $1.5 \le y_a/D_{50} \le 185$. To compare this range to the Limerinos equation range it may be noted that Limerinos also developed an equation for D_{50}. Implicit in a comparison of his equation for D_{84} and for D_{50} is a D_{84}/D_{50} ratio of 2.5. Using this ratio, the range of the Blodgett equation may be restated as $0.6 \le y_a/D_{84} \le 74$. This range is comparable to the range for Limerinos considering that y_a and R_h are close for the channels included in the study data.

Mussetter (1989) proposed a relationship that requires specification of bed material size, energy slope, and average depth.

$$n = \alpha \left(\frac{y_a}{D_{84}}\right)^{-0.46} \left(\frac{D_{84}}{D_{50}}\right)^{0.85} S_f^{0.39} \tag{C.4}$$

where,

n = Manning's roughness coefficient, dimensionless
y_a = average channel depth, ft (m)
S_f = friction (energy) slope, ft/ft (m/m)
α = 0.24

The Mussetter equation is applicable for $0.25 \le R_h/D_{84} \le 3.72$.

Bathurst (1981) provided a relationship for channels where individual rock elements protrude into the flow field significantly. This condition may be experienced on steep channels, but also occurs on moderate slopes.

$$n = \frac{\alpha \, y_a^{\frac{1}{6}}}{\sqrt{g} \; f(Fr) \; f(REG) \; f(CG)} \tag{C.5}$$

where,

y_a = average flow depth in the channel, ft (m)
g = acceleration due to gravity, 32.2 ft/s² (9.81 m/s²)
Fr = Froude number
REG = roughness element geometry
CG = channel geometry
α = unit conversion constant, 1.49 (CU), 1.0 (SI)

An appropriate equation selection must consider the basis on which the equation was developed and how it might apply within a closed conduit. The Bathurst, Jarrett, and Mussetter equations tend to better represent n values on steeper channels or channels with larger roughness elements. Limerinos and Blodgett attempt to encompass a wider range of conditions. The Bathurst equation depends on channel top width for calculation of Manning's n (See Kilgore and Cotton (2005) for details). However, in a closed conduit, top width does not monotonically increase with depth as it does in a natural channel. Therefore, the Bathurst equation would be problematic to apply within a culvert.

Appropriate selection should be based on the conditions being considered at the project site. Equations C.1, C.2, C.3, and C.4, as well as other relations documented in the literature, may be appropriate, but care should be taken in selecting a method and consideration should be given to performing a sensitivity analysis on the Manning's n because the methods do not always provide consistent results.

In addition, Equations C.1 through C.4 require either an average depth or hydraulic radius. In a natural channel both of these quantities can be considered to reflect the thickness of a layer of water on top of the bed. In a closed conduit, hydraulic radius retains this concept, however, the "bed" is the entire wetted perimeter, which includes culvert sidewalls in addition to the actual bed. Similarly, average depth is a function of top width, which monotonically increases with bed width in natural channels, but in many culverts decreases with increasing depth. Therefore, these quantities must be adapted for application inside a conduit. Maintaining the concept of a layer thickness over the bed, it is recommended that the water surface elevation minus the average bed elevation be taken as the depth for computing Manning's n rather than average depth or hydraulic radius.

C.3 SAND-BED CHANNELS

For sand-bed channels (D_{50} < 0.079 in (2 mm)), bedform influences the roughness of the channel. Evaluation of roughness for these channels is complex. The following guidance is a simplification of the more detailed description found in HDS 6 (Richardson, et al., 2001).

Bedforms are assessed by first considering whether the channel conditions are characteristic of an upper or lower flow regime. As shown in Figure C.1 (from Richardson, et al., 2001), the flow regime is determined based on the stream power and the D_{50} of the bed material. As shown in the figure, the lower flow regime may be characterized by dune, ripple, or plane bedforms. Between the lower and upper flow regimes is a transition zone.

The stream power is calculated as follows:

$$P = V\gamma y_a S_f \tag{C.6}$$

where,

P	=	stream power, ft-lb/s/ft^2, (N-m/s/m^2)
y_a	=	average flow depth in the channel, ft (m)
S_f	=	friction (energy) slope, ft/ft (m/m)
γ	=	specific weight of the water, 62.4 lb/ft^3 (9810 N/m^3)
V	=	average velocity, ft/s (m/s)

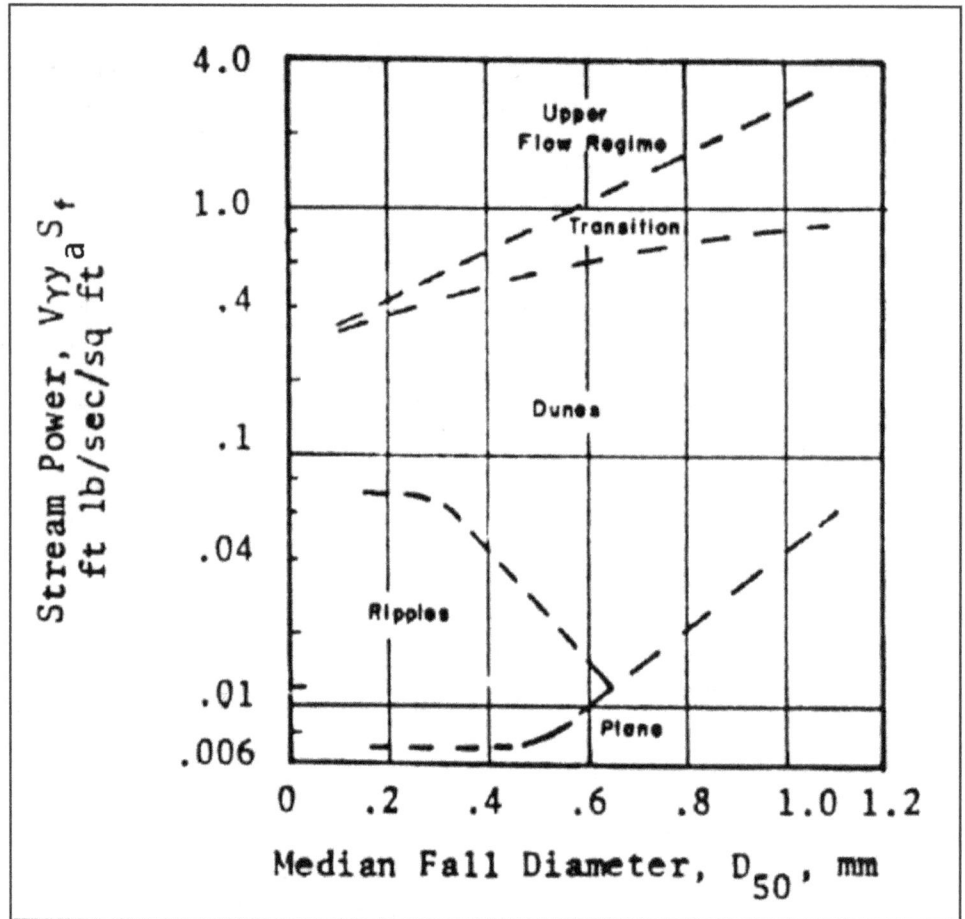

Figure C.1. Sand-bed Channel Flow Regime.

Within the various bedforms and regime classifications, a wide range of roughness values is possible. To simplify, the following procedure may be used:

1. For Q_P and Q_H assume upper flow regime and a Manning's n = 0.016. For Q_L assume lower flow regime and a Manning's n = 0.028.

2. Compute the stream power for all three flow conditions.

3. Based on the stream power and D_{50}, lookup the flow regime for each flow condition in Figure C.1.

4. If the flow regime from (3) matches the assumed flow regime in (1), the Manning's n values should be used. If not, repeat the analysis using the flow regime from (3).

C.4 COHESIVE SOILS

Cohesive soils are largely fine grained with a plasticity index greater than or equal to 10. A Manning's n of 0.016 may be used for these soils (Kilgore and Cotton, 2005).

C.5 FLOODPLAIN AND VEGETATED BANK ROUGHNESS

Roughness values may be required for floodplains, vegetated banks, and other natural stream characteristics beyond the bed and bank (non-vegetated) approaches previously given. Not intending to be exclusive, two are mentioned here. For additional information, refer to HDS 6 (Richardson, et al., 2001).

Barnes (1967) provides a useful reference with photos of a variety of natural streams and reports the estimated Manning's roughness for each stream. By comparing a stream of interest with those in Barnes' catalog, an estimate may be made.

Another general approach for estimating n values is summarized in Arcement and Schneider (1984) and consists of the selection of a base roughness value for a straight, uniform, smooth channel in the materials involved, then additive values are considered for the channel under consideration:

$$n = (n_0 + n_1 + n_2 + n_3 + n_4)m_5 \qquad (C.7)$$

where:

n_0	=	Base value for straight uniform channels
n_1	=	Additive value due to cross-section irregularity
n_2	=	Additive value due to variations of the channel
n_3	=	Additive value due to obstructions
n_4	=	Additive value due to vegetation
m_5	=	Multiplication factor due to sinuosity

Detailed values of the coefficients are found in Cowan (1956) and Arcement and Schneider (1984). It should not be assumed that all values of n_0 through n_4 are required to be non-zero. In fact, forcing all values to be non-zero may result in an overestimate of Manning's roughness.

See HDS 6 (Richardson, et al., 2001) for more information on these methods.

C.6 COMPOSITE ROUGHNESS VALUE

To analyze the culvert hydraulically, a composite Manning's n is required based on the combined effects of the bed and culvert wall roughness. Several approaches are available.

One approach is a simple weighting of n values based on wetted perimeter. This approach is used by the software tool FishXing and is stated as follows:

$$n_{comp} = \left[\frac{P_{bed}n_{bed} + P_{wall}n_{wall}}{P_{bed} + P_{wall}} \right] \qquad (C.8)$$

where,

n_{comp}	=	the composited n-value for the culvert.
P_{bed}	=	wetted perimeter of the natural material in the culvert.
n_{bed}	=	n-value of the natural material in the culvert.
P_{wall}	=	wetted perimeter of the culvert walls above the natural material.
n_{wall}	=	n-value of the culvert material.

Another alternative, applied by the software tool HEC-RAS and described in HDS 5 (Normann, et al., 2005), is Equation C.9.

$$n_{comp} = \left[\frac{P_{bed}n_{bed}^{1.5} + P_{wall}n_{wall}^{1.5}}{P_{bed} + P_{wall}} \right]^{2/3}$$ (C.9)

where,

n_{comp} = the composited n-value for the culvert.

P_{bed} = wetted perimeter of the natural material in the culvert.

n_{bed} = n-value of the natural material in the culvert.

P_{wall} = wetted perimeter of the culvert walls above the natural material.

n_{wall} = n-value of the culvert material.

Both of the approaches represented in Equations C.8 and C.9 assume that wetted perimeter is an appropriate weighting parameter because the wetted perimeter represents the surface over which the roughness is applied. However, the energy lost to friction is a function of the roughness (which is represented by Manning's n) and the boundary force applied to the boundary. The boundary force may be represented as follows:

$$F = \tau P L$$ (C.10)

where,

F = boundary force on the wetted perimeter, lb (N)

τ = shear stress on the conduit boundary, lb/ft^2 (N/m^2)

P = wetted perimeter, ft (m)

L = conduit length, ft (m)

Substituting the relationship for boundary shear stress, $\tau = \gamma y S$, Equation C.10 becomes:

$$F = \gamma y S P L$$ (C.11)

For a prismatic shape, the values for γ, S, and L are constants leaving the boundary force proportional to depth and wetted perimeter:

$$F \propto y P$$ (C.12)

Equations C.8 and C.9 apply a weighting based on wetted perimeter, P, and essentially assume that an average depth across the channel is a reasonable assumption. However, in closed conduits, the depth of water on the conduit walls is not well represented by an average depth. Therefore, an alternative weighting is proposed.

In a culvert, the bed experiences the full depth of flow from a shear stress perspective and the wall experiences an average depth between zero and the full depth. For a box culvert, the average depth is one-half the full depth. For other culvert shapes, the average depth at the wall depends on the size, and shape of the culvert. Equation C.13 incorporates this concept by introducing a wall shape coefficient in the weighting:

$$n_{comp} = \left[\frac{P_{bed}n_{bed} + c_y P_{wall}n_{wall}}{P_{bed} + c_y P_{wall}} \right]$$ (C.13)

where,

c_y = wall shape coefficient, between 0 and 1.

The wall shape coefficient is 0.5 for a box culvert. As an approximation, 0.5 may be used for other shapes.

Figure C.2 summarizes sample computations of a composite n value for a 5' x 5' (1.5 m x 1.5 m) concrete box culvert (n = 0.013) with an embedment of 1 ft (0.3 m) with material having D_{84} = 0.1 ft (0.03 m). Manning's n is calculated for the bed material using the Limerinos equation. The solid lines in the figure result from Equations C.8, C.9, and C.13 using the hydraulic radius, R_h, in the Limerinos equation (Equation C.1). All of the equations show a decrease in Manning's roughness with depth, as more of the sidewall is wetted.

Figure C.2 also shows the use of the same three equations, but with the use of depth, y, in the culvert in the Limerinos equation (Equation C.1). For shallow depths, use of depth rather than hydraulic radius has a much greater effect on the composite n than does the choice of compositing approach. However, as depth increases, the estimates converge.

Composite n may be calculated from any of the three methods described in this section. As stated earlier, however, the depth, y, rather than average depth or hydraulic radius should be used in the estimate of Manning's n for the bed material in the culvert.

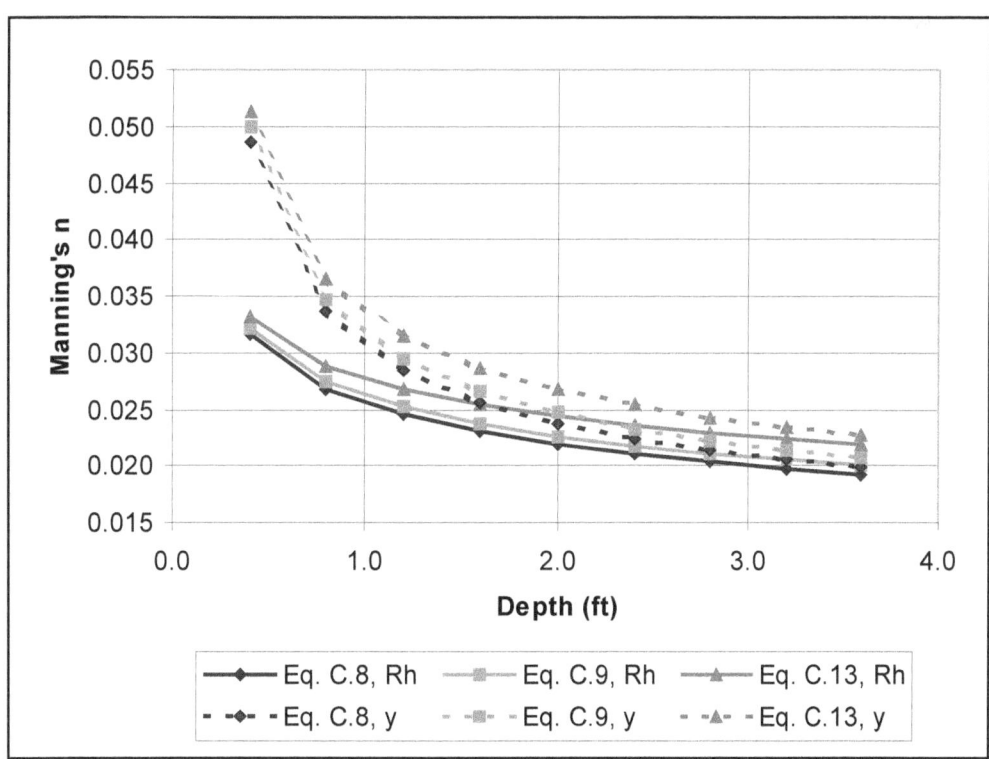

Figure C.2. Alternative Estimates for Composite n Values.

This page intentionally left blank.

APPENDIX D- PERMISSIBLE SHEAR STRESS

Bed material is considered stable if the boundaries are basically immobile (static equilibrium). A fundamental tool for assessing this condition is the permissible tractive force (shear stress) approach. The tractive force (boundary shear stress) approach focuses on stresses developed at the interface between flowing water and materials forming the channel boundary. By Chow's definition, permissible tractive force is the maximum unit tractive force that will not cause serious erosion of channel bed material from a level channel bed (Chow, 1959).

D.1 NONCOHESIVE SOILS

D.1.1 Shield's Equation

Shield's Equation is appropriate for general sediment transport analyses as it has been shown that it defines particle instability as the point where *general* movement of that particle size is occurring. It is probably the most-widely used and most-widely accepted permissible shear stress determination method:

$$F_* = \frac{\tau_o}{(\gamma_s - \gamma)d} \tag{D.1}$$

where,

F_*	=	Shield's parameter, dimensionless
τ_o	=	shear stress at which movement begins, lb/ft^2 (N/m^2)
γ_s	=	specific weight of the stone, lb/ft^3 (N/m^3)
γ	=	specific weight of the water, 62.4 lb/ft^3 (9810 N/m^3)
d	=	particle size, ft (m)

The form of the Shield's equation for determining permissible shear stress is obtained by reworking Equation D.1 as follows:

$$\tau_p = F_*(\gamma_s - \gamma)d \tag{D.2}$$

where,

τ_p = permissible shear stress, lb/ft^2 (N/m^2)

Equation D.2 is valid for slopes up to 10 percent and is based on assumptions related to the relative importance of skin friction, form drag, and channel slope. However, skin friction and form drag have been documented to vary resulting in reports of variations in Shield's parameter by different investigators, for example Gessler (1965), Wang and Shen (1985), and Kilgore and Young (1993). This variation is usually linked to particle Reynolds number as defined below:

$$R_e = \frac{V_* d}{\nu} \tag{D.3}$$

where,

R_e	=	particle Reynolds number, dimensionless
V_*	=	shear velocity, ft/s (m/s)
ν	=	kinematic viscosity, 1.217x10^{-5} ft^2/s at 60 deg F (1.131x10^{-6} m^2/s at 15.5 deg C)

Shear velocity is defined as:

$$V_* = \sqrt{gyS}$$ (D.4)

where,

g = gravitational acceleration, 32.2 ft/s² (9.81 m/s²)
y = maximum channel depth, ft (m)
S = channel slope, ft/ft (m/m)

The variation in Shield's parameter with Reynolds number is summarized in Table D.1. For particle Reynolds numbers is less than 4x10⁴, a common situation, a Shields parameter of 0.047 is appropriate. For cases with a particle Reynolds number greater than 2x10⁵, such as may be found in channels on steeper slopes, a higher Shields parameter of 0.15 is appropriate because of the changes in relative importance of skin friction and form drag as noted earlier. For Reynolds numbers in between these two values, Shields parameter should be interpolated based on the Reynolds number.

Table D.1.   Selection of Shields' Parameter.

Reynolds number	Shield's Parameter	Safety Factor (SF)	Effective Shield's Parameter
≤ 4x10⁴	0.047	1.0	0.047
4x10⁴<R$_e$<2x10⁵	Linear interpolation	Linear interpolation	Linear interpolation
≥ 2x10⁵	0.15	1.5	0.10

Source: From Kilgore and Cotton (2005)

For channel flow conditions with higher Reynolds numbers there is often uncertainty in estimating Manning's roughness, depths, and velocities. In the context of riprap revetment design, Kilgore and Cotton (2005) used a sliding safety factor to address this uncertainty that increases with Reynolds number. It is recommended that these safety factors, as summarized in Table D.1, are also applicable in the context of this manual. The final column in Table D.1 provides the effective Shield's parameter computed as the Shield's parameter divided by the safety factor. The effective Shield's parameter is used in Equation D.2.

D.1.2 Modified Shield's Equation

As described in FSSWG (2008), the modified critical shear stress equation is based on the relationship between the particle size of interest, D_i, and D_{50}, which is assumed to be unaffected by the shielding/exposure effect (Andrews, 1983, Bathurst, 1987; Komar, 1987; Komar, 1996; Komar and Carling, 1991). For the particle size of interest, D_i, for example D_{84}, when this larger particle size begins to move, much of the streambed is in motion and the structure of the channel bed will change.

The modified permissible shear stress equation (Komar 1987; Komar 1996; Komar and Carling, 1991) is as follows:

$$\tau_p = F_* (\gamma_s - \gamma) D_i^{0.3} D_{50}^{0.7}$$ (D.5)

where,

τ_p = permissible shear stress, lb/ft^2 (N/m^2)

F_* = Shields parameter for D_{50} particle size (this value is obtained from Table D.1)

γ_s = specific weight of the stone, lb/ft^3 (N/m^3)

γ = specific weight of the water, 62.4 lb/ft^3 (9810 N/m^3)

D_i = particle size of interest, ft (m)

The modified critical shear stress equation is appropriate for assessing particle stability in riffles and plane-bed channels (i.e., where flow is relatively uniform or gradually varied between cross sections) with channel-bed gradients less than 5 percent and D_{84} particles ranging between 2.5 to 10 inches (10 and 250 mm). In addition, the diameter for the particle size of interest (e.g., D_{84}) must not be larger than 20 to 30 times the D_{50} particle diameter. For D_i/D_{50} ratios greater than 30, Equation D.5 is not applicable because a large particle will roll easily over surrounding smaller sediments (Komar, 1987, 1996; Carling, 1992). D_{84}/D_{50} or D_{95}/D_{50} ratios are typically less than 5 in natural channels.

D.1.3 Fine-grained Noncohesive Soils

The permissible shear stress for fine-grained, non-cohesive soils ($D_{75} < 0.05$ in (1.3 mm)) is relatively constant and is conservatively estimated to be 0.02 lb/ft^2 (1.0 N/m^2) (Kilgore and Cotton, 2005).

D.2 COHESIVE SOILS

Cohesive soils are largely fine grained and their permissible shear stress depends on cohesive strength and soil density. Cohesive strength is associated with the plasticity index (PI), which is the difference between the liquid and plastic limits of the soil. The soil density is a function of the void ratio (e). The basic formula for permissible shear on cohesive soils is the following.

$$\tau_{p,soil} = \left(c_1 PI^2 + c_2 PI + c_3\right)\left(c_4 + c_5 e\right)^2 c_6 \qquad (D.6)$$

where,

$\tau_{p,soil}$ = soil permissible shear stress, lb/ft^2 (N/m^2)

PI = plasticity index

e = void ratio

c_1, c_2, c_3, c_4, c_5, c_6 = coefficients from Table D.2, (USDA, 1987)

A simplified approach for estimating permissible soil shear stress based on Equation D.6 is illustrated in Figure D.1 (Kilgore and Cotton, 2005). Fine-grained soils are grouped together (GM, CL, SC, ML, SM, and MH) and coarse-grained soil (GC). Clays (CH) fall between the two groups.

Higher soil unit weight increases the permissible shear stress and lower soil unit weight decreases permissible shear stress. Figure D.1 is applicable for soils that are within 5 percent of a typical unit weight for a soil class. For sands and gravels (SM, SC, GM, GC) typical soil unit weight is approximately 100 lb/ft^3 (1.6 ton/m^3), for silts and lean clays (ML, CL) 90 lb/ft^3 (1.4 ton/m^3) and fat clays (CH, MH) 80 lb/ft^3 (1.3 ton/m^3).

Table D.2. Coefficients for Permissible Soil Shear Stress.

ASTM Soil Classification[1]	Applicable Range	c_1	c_2	c_3	c_4	c_5	c_6 (SI)	c_6 (CU)
GM	$10 \leq PI \leq 20$	1.07	14.3	47.7	1.42	-0.61	4.8×10^{-3}	10^{-4}
	$20 \leq PI$			0.076	1.42	-0.61	48.	1.0
GC	$10 \leq PI \leq 20$	0.0477	2.86	42.9	1.42	-0.61	4.8×10^{-2}	10^{-3}
	$20 \leq PI$			0.119	1.42	-0.61	48.	1.0
SM	$10 \leq PI \leq 20$	1.07	7.15	11.9	1.42	-0.61	4.8×10^{-3}	10^{-4}
	$20 \leq PI$			0.058	1.42	-0.61	48.	1.0
SC	$10 \leq PI \leq 20$	1.07	14.3	47.7	1.42	-0.61	4.8×10^{-3}	10^{-4}
	$20 \leq PI$			0.076	1.42	-0.61	48.	1.0
ML	$10 \leq PI \leq 20$	1.07	7.15	11.9	1.48	-0.57	4.8×10^{-3}	10^{-4}
	$20 \leq PI$			0.058	1.48	-0.57	48.	1.0
CL	$10 \leq PI \leq 20$	1.07	14.3	47.7	1.48	-0.57	4.8×10^{-3}	10^{-4}
	$20 \leq PI$			0.076	1.48	-0.57	48.	1.0
MH	$10 \leq PI \leq 20$	0.0477	1.43	10.7	1.38	-0.373	4.8×10^{-2}	10^{-3}
	$20 \leq PI$			0.058	1.38	-0.373	48.	1.0
CH	$20 \leq PI$			0.097	1.38	-0.373	48.	1.0

(1) Note: Typical names

GM   Silty gravels, gravel-sand silt mixtures

GC   Clayey gravels, gravel-sand-clay mixtures

SM   Silty sands, sand-silt mixtures

SC   Clayey sands, sand-clay mixtures

ML   Inorganic silts, very fine sands, rock flour, silty or clayey fine sands

CL   Inorganic clays of low to medium plasticity, gravelly clays, sandy clays, silty clays, lean clays

MH   Inorganic silts, micaceous or diatomaceous fine sands or silts, elastic silts

CH   Inorganic clays of high plasticity, fat clays

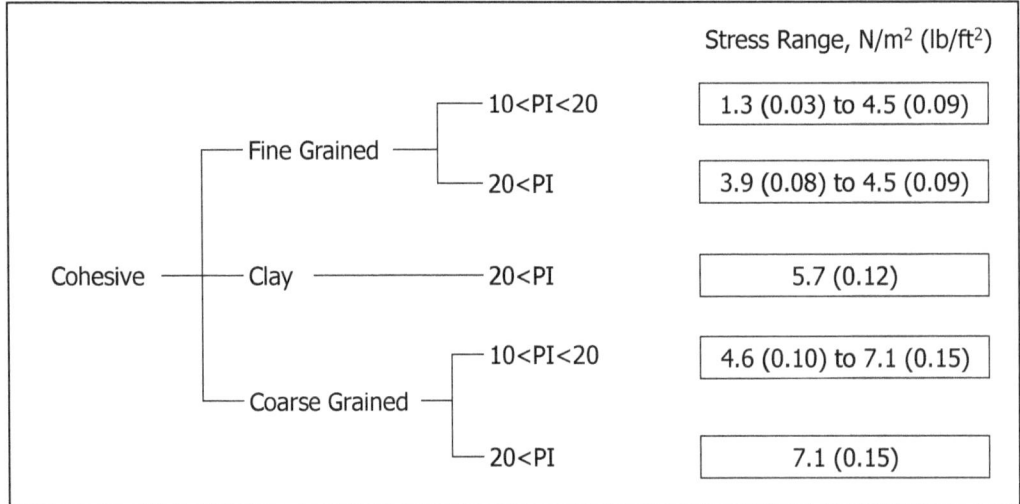

Figure D.1. Cohesive Soil Permissible Shear Stress.

One thing to note is that the critical shear stress for cohesive sediment *deposition* is less (possibly much less) than the shear stress required for cohesive sediment *erosion*-particularly if there has been a sufficient amount of time in between events (i.e. the fine sediment has had time to consolidate and bind). It is quite possible that a culvert could be designed to withstand cohesive sediment erosion at high flow but allow deposition at low flow. Over time, this could lead to clogging of the culvert. It may be appropriate for the design to be such that deposition of fine particles is prevented, even at low flows. In this case, artificial anchoring of coarser sediments may be necessary to provide a constant natural bed.

D.3 CRITICAL UNIT DISCHARGE

An alternative approach to shear stress is provided by a critical unit discharge approach. The following text is taken from FSSWG (2008).

For channels steeper than 1 percent (S = 0.01) where the flow depth is shallow with respect to the channel bed particle sizes ($R/D_{50} < 10$), water depth can be quite variable because large rocks or wood pieces on or near the surface influence depth (Bathurst 1987). For such channels, Bathurst et al. (1987) used flume data to construct the following equation, which predicts the critical unit discharge for entraining the D_{50} particle size in well-sorted sediments:

$$q_{c-D50} = \frac{0.15g^{0.5}D_{50}^{1.5}}{S^{1.12}}$$

(D.7)

where,

q_{c-D50} = critical unit discharge to entrain the D_{50} particle size, ft³/ft (m³/m)

D_{50} = median or 50th percentile particle size, ft (m)

g = gravitational acceleration, 32.2 ft/s² (9.8 m/s²)

S = bed slope, ft/ft (m/m)

In the flume studies, particle sizes ranged between 0.1 and 1.7 in (3 and 44 mm), the experimental bed materials were uniform (i.e., well-sorted), slopes ranged between 0.0025 and 0.20 ft/ft (m/m), and ratios of water depth to particle size approached 1 (Bathurst, 1987).

Bathurst (1987) used Equation D.7 to predict the entrainment of particles in poorly sorted channel beds, by comparing the particle size of interest (e.g., D_{84} or D_{95}) to a reference particle size. The reference particle size is the D_{50} particle size, which is assumed to move at the same flow as in a well-sorted channel. The critical unit discharge for entraining a particle size of interest is determined by:

$$q_{ci} = q_{c-D50}\left(D_i / D_{50}\right)^b$$

(D.8)

where,

q_{ci} = critical unit discharge to entrain the particle size of interest, ft³/ft (m³/m)

D_i = particle size of interest, in (mm)

D_{50} = median or 50th percentile particle size, in (mm)

The exponent b is a measure of the range of particle sizes that make up the channel bed. It quantifies the effects on particle entrainment of smaller particles being hidden and of larger particles being exposed to flow. Calculate the exponent from:

$$b = 1.5\left(D_{16} / D_{84}\right)$$

(D.9)

where,

D_{84} = 84th percentile particle size, in (mm)

D_{16} = 16th percentile particle size, in (mm)

Equations D.8 and D.9 were derived from limited data and are most appropriate for assessing particle stability in riffles and plane-bed channels (i.e., where flow is relatively uniform or gradually varied between cross sections) with slopes ranging between 0.036 and 0.052 ft/ft (m/m), widths ranging between 20 and 36 ft (6.1 and 11.0 m), D_{16} particle sizes between 1.3 and 2.3 in (32 and 58 mm), D_{50} particle sizes between 2.8 and 5.5 in (72 and 140 mm), and D_{84} particle sizes between 6 and 10 in (156 and 251 millimeters).

APPENDIX E- EMBEDMENT

A variety of Federal and state agencies have specified embedment criteria. Unfortunately, the criteria are generally stated without an explanation of the underlying rationale. Table E.1 summarizes several agency embedment criteria.

Table E.1. Summary of Culvert Embedment Criteria.

Agency/Procedure	Criteria
Alaska Dept. of Fish and Game and Dept. of Transportation/Stream Simulation (Alaska Dept. of Fish and Game and Dept. of Transportation, 2001)	Circular pipes: 40 percent of the rise. Pipe arches: 20 percent of the rise.
California Department of Transportation/Active Channel (CALTRANS, 2007)	30 – 50 percent of culvert rise.
Western Federal Lands Division, Federal Highway Administration/Browning (Browning, 1990)	Culvert rise less than 10 ft (3.2 m) diameter: minimum of 1-2 ft (0.3-0.6 m). Culvert rise greater than or equal to 10 ft (3.2 m): minimum of 1/5th the culvert rise. Where system wide degradation is possible, installation may require additional embedment to match the anticipated stream surface lowering.
Maine/Hydraulic Design (Maine Department of Transportation, 2004)	Culvert diameter less than or equal to 4 ft (1.22 m): 0.5 ft (0.15 m). Culvert diameter greater than 4 ft (1.22 m): 1 ft (0.30 m). Embedment is allowed to fill naturally after culvert placement.
Maryland/General (Maryland State Highway Administration, 2005)	20 percent of culvert rise. Embedment is allowed to fill naturally after culvert placement.
Oregon Department of Fish and Wildlife/Stream Simulation (Robison, et al., 1999)	Circular culvert: maximum of 40 percent of rise or 2 ft (0.6 m). Pipe-arch or box culvert: maximum of 20 percent of rise or 1.5 ft (0.46 m).
Vermont/Stream Simulation (Bates and Kirn, 2007)	Maximum of 1.5 times the diameter of the largest immobile particles in the bed or 4 times the size of the largest mobile material
Washington Department of Fish and Wildlife/Roughened Channel (Bates, et al., 2003)	Circular: 30 percent of rise
Washington Department of Fish and Wildlife/Hydraulic Design (Bates, et al., 2003)	20 percent of culvert rise.
Washington Department of Fish and Wildlife/No Slope and Active Channel (Bates, et al., 2003)	20 percent of rise at outlet, no more than 40 percent at inlet.
Washington Department of Fish and Wildlife/Stream Simulation (Bates, et al., 2003)	30-50 percent of culvert rise.

With the exception of the Vermont criterion in Table E.1, other agency criteria overlook bed material as being a potential limiting factor for embedment depth. This concept has been applied for many years in the design of rock linings and aprons.

This page intentionally left blank.

APPENDIX F- BED GRADATION

Generally, it is desirable for bed gradation in the culvert to be similar to the gradation in the natural stream. HDS 6 (Richardson, et al., 2001) describes a gradation coefficient. A target gradation coefficient may be defined and the necessary quantiles developed from Equation F.1.

$$G = \frac{1}{2}\left[\frac{D_{50}}{D_{16}} + \frac{D_{84}}{D_{50}}\right] \tag{F.1}$$

where,

G = gradation coefficient

D_x = sediment diameter particle of which x percent of the sample, by weight, is finer

A second method for creating a well-graded bed mixture based on larger size fractions from a pebble count or from armored streambeds is the equation developed by Fuller and Thompson (1907), which defines dense sediment mixtures commonly used by the aggregate industry. This equation has not yet been widely field-tested for this application, so good professional judgment is critical (USFS, 2008).

The Fuller-Thompson equation is:

$$P = \left(\frac{D_x}{D_{max}}\right)^m \tag{F.2}$$

where,

D_x = any particle size of interest

P = fraction of the mixture smaller than D_x

D_{max} = the largest size material in the mix

m = parameter that determines how fine or coarse the resulting mix will be. A value of 0.5 produces a maximum density mix when particles are round.

The Fuller-Thompson equation can be rearranged to base the particle size determination on any size fraction. Using D_{50} as the reference size, the equations for D_{95}, D_{84}, D_{16} and D_5 are:

$$D_{95} = (1.9)^{1/m} D_{50} \tag{F.3a}$$

$$D_{84} = (1.68)^{1/m} D_{50} \tag{F.3b}$$

$$D_{16} = (0.32)^{1/m} D_{50} \tag{F.3c}$$

$$D_5 = (0.1)^{1/m} D_{50} \tag{F.3d}$$

To develop the particle-size distribution curve for the culvert bed mix, use m values between 0.45 and 0.70, a standard range for high-density mixes (FSSWG, 2008). Using the Fuller-Thompson method does not necessarily reproduce the natural subsurface particle size distribution in the adjacent streambed, but it does result in a dense, well-graded distribution.

Similar results may be obtained by smoothly redrawing the lower half of the particle size distribution curve by hand, such that the tail has an appropriate percentage of fines smaller than 0.079 in (2 mm) (FSSWG, 2008).

These design procedures may result in a bed mix that is coarser overall than the adjacent reach subsurface gradation resulting in a conservative design. If the bed scours, there will be additional armor material below the surface, and the resulting bed surface will become coarser and rougher.

APPENDIX G-   BAFFLES AND SILLS

Baffles are a series of evenly spaced vertical extensions attached to the culvert bottom. Culvert baffles can be used to meet a variety of design objectives:

- Dissipating energy.

- Providing velocity diversity for AOP.

- Providing grade control.

- Preventing interstitial flow in the culvert bed material.

Many studies have been completed to determine the effects of flow velocities and turbulence on baffle design, e.g. Thurman and Horner-Devine (2007), particularly for achieving the first two objectives. However, these studies do not consider the presence of bed material because the devices evaluated are not intended to operate with retained bed material. Therefore, design approaches for baffles for bed retention are lacking and they should not be relied on for bed retention until research documenting effectiveness is conducted. However, baffles have been used successfully to retain larger materials, such as riprap in the bottom of a culvert or channel.

Until field experience or research provides alternative guidance, baffles used for objectives other than bed retention should be placed across the culvert bottom with the top of the baffle at the elevation of the proposed bed. If a low-flow channel is required, this should be provided for in the baffle profile. For example, if a triangular low-flow channel is desired, the baffle profile should be the same dimensions as the triangular low-flow channel. Baffle spacing likely depends on the discharge conditions and culvert slope.   Bates and Love (2009) provide a good overview of baffle configuration and design.

Sills are essentially a singular baffle located at the culvert outlet. A sill can be used both for grade control and to prevent interstitial flow in the culvert bed material.

This page intentionally left blank.

APPENDIX H- DESIGN EXAMPLE: NORTH THOMPSON CREEK, COLORADO

H.1 SITE DESCRIPTION

The design procedure is applied to a road crossing of the North Thompson Creek, which is approximately 15 miles (24 kilometers) south and 3 miles (4.8 kilometers) east of Glenwood Springs, Colorado. (See Figure H.1.) The drainage area to the crossing is 2.33 mi^2 (6.03 km^2). The watershed is forested (spruce, fir, and aspen) with areas of meadow. Activities in the watershed include logging, cattle grazing, and recreational use. Elevations in the watershed range from 9,400 to 10,910 ft (2865 to 3325 m). The Wasach and Ohio Creek formations make up the geology with interbedded sandstones and shales mixed with coarse basalt remnants. (All data and photos for this application were provided by Mark Weinhold of the USFS.)

There is an existing 36-in (910-mm) culvert at the stream-road crossing. Figures H.2a and H.2b show the inlet and outlet of the culvert, respectively. The culvert was identified for replacement by the USFS because it was considered a passage barrier, possibly because of high velocities in the barrel.

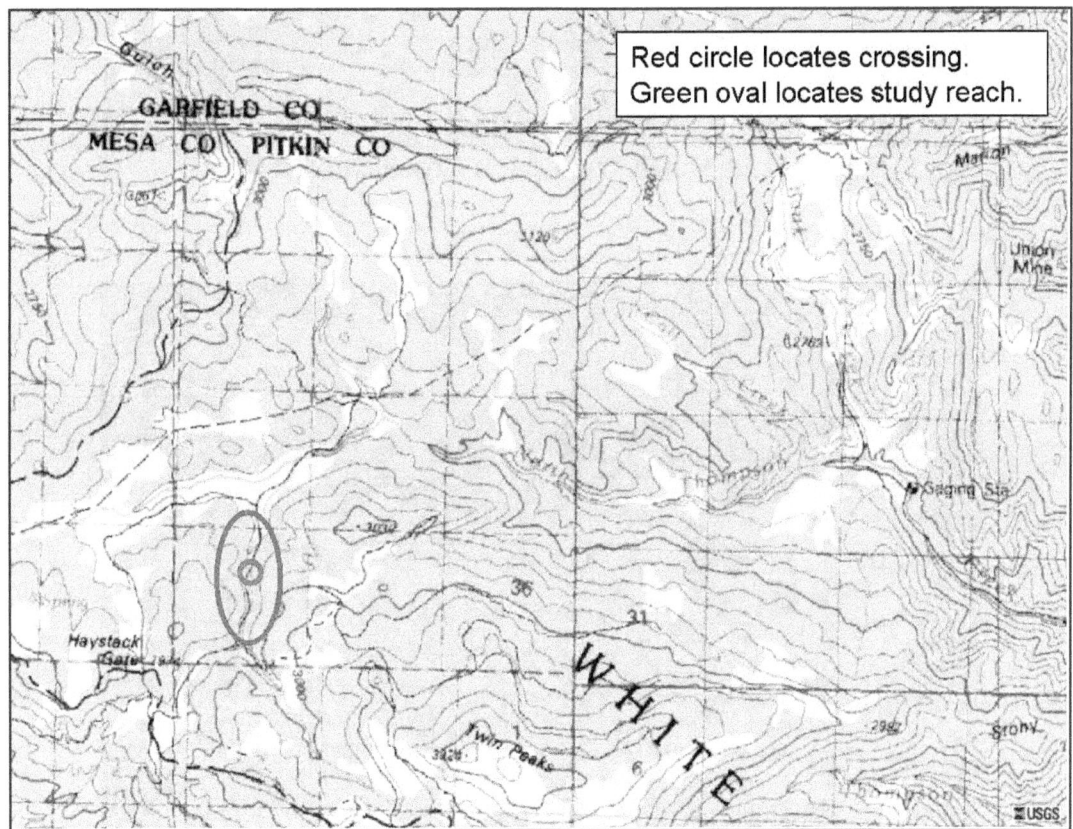

Figure H.1. North Thompson Creek.

Figure H.2a. North Thompson Creek Culvert Inlet.

Figure H.2b. North Thompson Creek Culvert Outlet.

H.2 DESIGN PROCEDURE APPLICATION

This section illustrates the application of the design procedure. The uses of two separate tool sets are shown: 1) HY-8 with normal depth computations for the channel cross-sections and 2) HEC-RAS. Although both tool sets are shown, the designer may choose one or the other as appropriate for the site and the designers modeling skills.

Step 1. Determine Design Flows.

Discharges are determined for the peak flow, Q_P, high passage flow, Q_H, and low passage flow, Q_L. The drainage area was delineated using the Watershed Modeling System (WMS) and is shown in Figure H.3. Watershed characteristics computed by WMS are summarized in Table H.1.

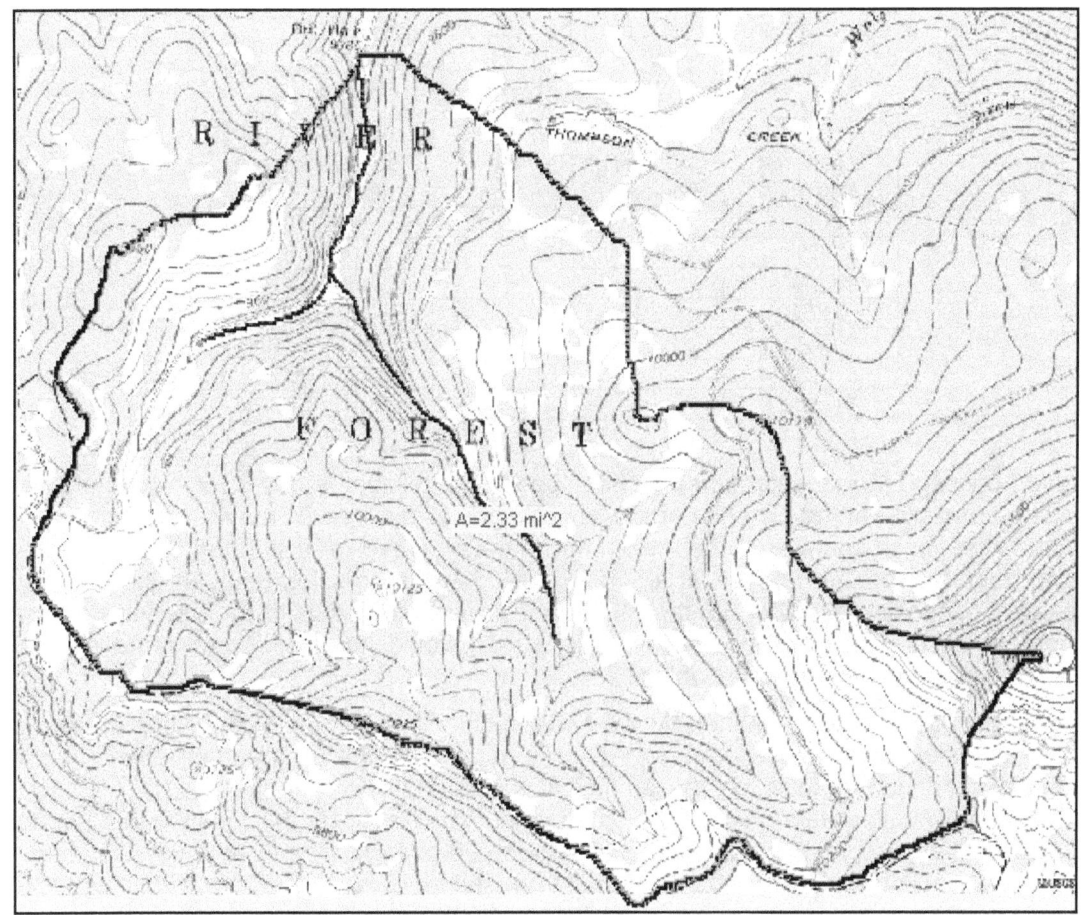

Figure H.3. North Thompson Creek Drainage Area Delineation.

Table H.1. Watershed and Rainfall Characteristics.

Characteristic	CU	SI
Drainage Area	2.33 mi^2	6.03 km^2
Mean Basin Slope	0.19 ft/ft	0.19 m/m
Mean Basin Elevation	9915 ft	3022 m
Mean Areal Precipitation	34 in	864 mm

The FHWA Central Federal Lands Highway Division (CFLHD) and Colorado Department of Transportation (CDOT) standards for culvert design are applied for this site. The CFLHD standard for culverts under a low-standard road is a 25-yr return period. The CDOT standard for cross drainage on rural two-lane roads varies with stream size and average daily traffic (ADT). For small streams (Q_{50} less than 4,000 ft^3/s (113 m^3/s)) and ADT greater than 750, the standard is the 25-yr frequency. Therefore, both CFLHD and CDOT standards call for the 25-yr event.

Two USGS regression equation sources are considered and summarized in Table H.2 with the corresponding 25-yr discharges. The 25-yr discharge adopted for this design is the Blakemore et al. (1997) value of 103 ft^3/s (2.92 m^3/s). In addition to being the most conservative, it is based on more recent equations than the Kircher et al. (1985) equations. Another source, Vaill (2000), provides more recent equations, however, these equations apply to basins greater than 5.5 mi^2 (14 km^2). The subject basin area is less than half this lower limit.

Table H.2. Discharge Estimates.

Discharge Quantity	Blakemore, et al. (1997) High Elevation Region 1, ft^3/s (m^3/s)	Kircher, et al. (1985) Mountain Region, ft^3/s (m^3/s)
Q_{25}	Q_P = 103 (2.92)	66 (1.9)
Q_2	41 (1.2)	30 (0.85)
$Q_{10\%}$	--	Q_H = 8.8 (0.25)
$0.25Q_2$	10.2 (0.3)	7.5 (0.21)
$Q_{90\%}$	--	0.15 (0.0042)
7Q2	--	0.13 (0.0037)
Q_L (min)		Q_L = 1 (0.028)

The high passage flow is determined by site-specific guidelines, if they exist. None are known to exist for this site. In the absence of site-specific guidelines, the Q_H may be defined as the 10 percent exceedance quantile on the annual flow duration curve. A flow duration curve does not exist for this location, but Kircher, et al. (1985) includes a regression equation that results in a 10 percent exceedance flow of 8.8 ft^3/s (0.25 m^3/s), which will be used for Q_H in this design. If an appropriate equation had not been available, Q_H would have been estimated as $0.25Q_2$. These values are summarized in Table H.2.

The low passage flow is determined by site-specific guidelines, if they exist. None are known to exist for this site. In the absence of site-specific guidelines, the Q_L may be defined as the 90 percent exceedance quantile on the annual flow duration curve or the 7-day, 2-yr low flow (7Q2). As previously noted, a flow duration curve does not exist for this location, but Kircher, et al. (1985) includes regression equations for both the 90 percent exceedance flow and the 7Q2, which result in flows of 0.15 ft^3/s (0.0042 m^3/s) and 0.13 ft^3/s (0.0037 m^3/s), respectively. However, both are less than the minimum low passage flow of 1 ft^3/s (0.028 m^3/s), so 1 ft^3/s (0.028 m^3/s) will be used for Q_L in this design. Low passage values are summarized in Table H.2.

Step 2. Determine Project Reach and Representative Channel Characteristics.

The project reach should, at a minimum, extend no less than three culvert lengths or 200 ft (61 m), whichever is greater, up and downstream of the crossing location. Since the existing culvert is 46 ft (14 m) in length, the project reach must extend at least 200 ft (61 m) upstream and downstream of the culvert inlet and outlet, respectively. At least three cross-sections should be obtained both upstream and downstream from the crossing location.

Nine stream cross-sections were collected by the USFS. Four are downstream of the culvert; the most downstream cross-section is approximately 200 ft (61 m) downstream from the road centerline. Five are upstream; the most upstream cross-section is approximately 300 ft (91 m) upstream of the road centerline. Table H.3 summarizes the cross-section locations and Figure H.4 shows the creek and cross-sections schematically in plan view. Cross-sections shown in Figure H.4, but not listed in Table H.3 were interpolated for the purpose of water surface profile modeling with HEC-RAS. Plots of the surveyed cross-sections are included in H.3.1.

Table H.3.   Surveyed Cross-Sections.

Cross-section	Station (ft)	Station (m)	Thalweg Elevation (ft)	Thalweg Elevation (m)	Slope to downstream cross-section (ft/ft or m/m))
567	567.0	172.8	101.83	31.04	0.032
472	471.7	143.8	98.79	30.11	0.024
399	399.0	121.6	97.05	29.58	0.011
342	341.7	104.2	96.44	29.39	0.006
307	306.5	93.4	96.21	29.32	0.043
Road centerline	260	79.2	--	--	--
215	215.2	65.6	92.26	28.12	0.038
172	171.6	52.3	90.59	27.61	0.016
125	124.7	38.0	89.83	27.38	0.024
57	56.8	17.3	88.20	26.88	0.026

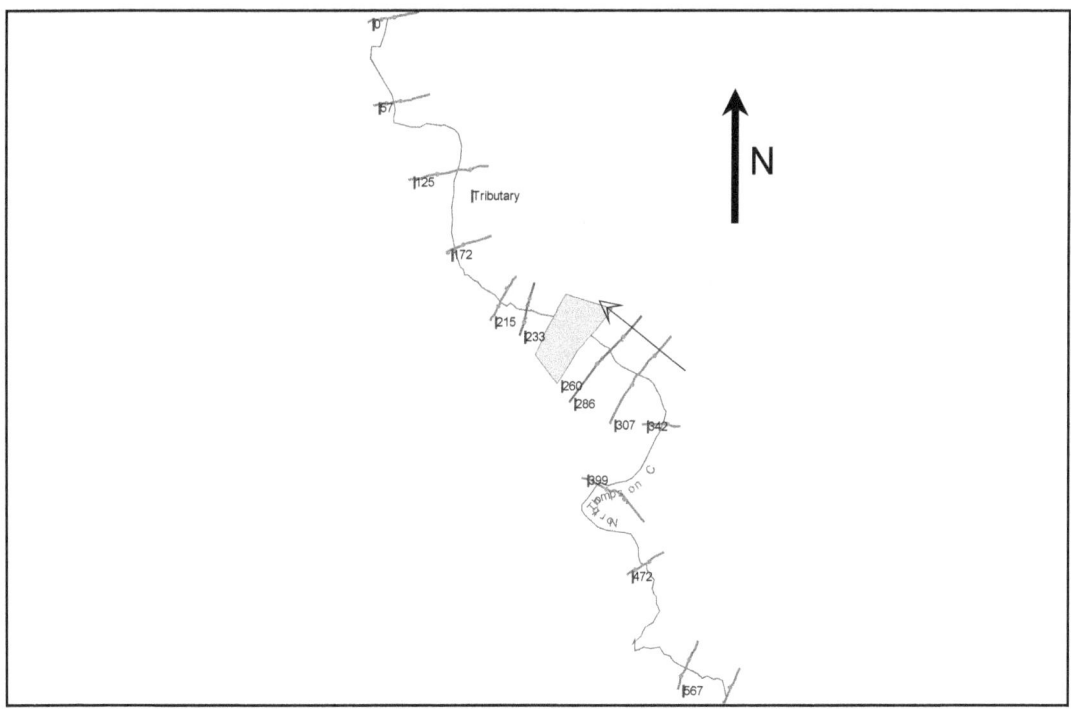

Figure H.4.   Creek and Cross-section Schematic.

The longitudinal profile of the stream and existing roadway embankment is shown in Figure H.5 using all surveyed data collected by the USFS. Although only the data acquired at the cross-section locations are used, the longitudinal detail shows the variability within the natural stream. Superimposed on the detailed profile are the cross-section locations plotted with their thalweg elevations as well as the existing culvert invert. The existing culvert slope is 3.95 percent. The profile shows deposition of material at the culvert inlet and a scour hole at the culvert outlet.

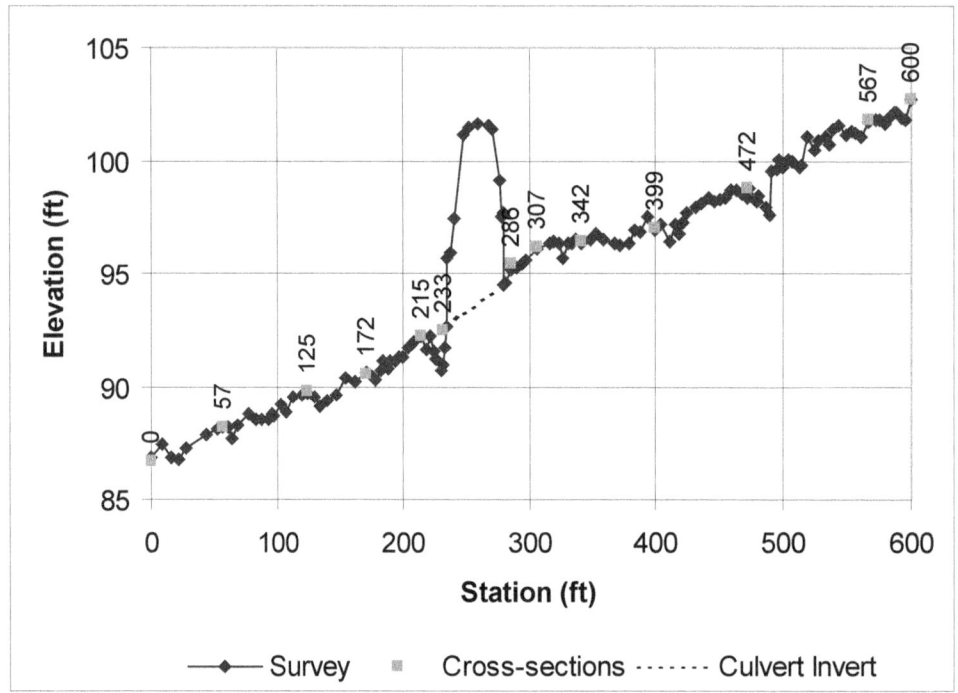

Figure H.5. Longitudinal Profile.

Two bed material gradations were collected in the project reach: 1) at cross-section 57 and 2) between cross-sections 172 and 215 in a riffle. Pebble counts were used to determine the gradations as summarized in Table H.4 and Figure H.6. Evidence of bed armoring was not reported. The unit weight of the bed material was not provided so a value of 156 lb/ft^3 (24,500 N/m^3) is assumed.

Table H.4. Bed Material Quantiles.

Quantile	XS 57		XS 172/215	
	Size (ft)	Size (mm)	Size (ft)	Size (mm)
D_{95}	0.715	218	0.935	285
D_{84}	0.495	151	0.636	194
D_{50}	0.180	55	0.148	45
D_{16}	0.072	22	0.069	21
D_5	0.043	13	0.043	13

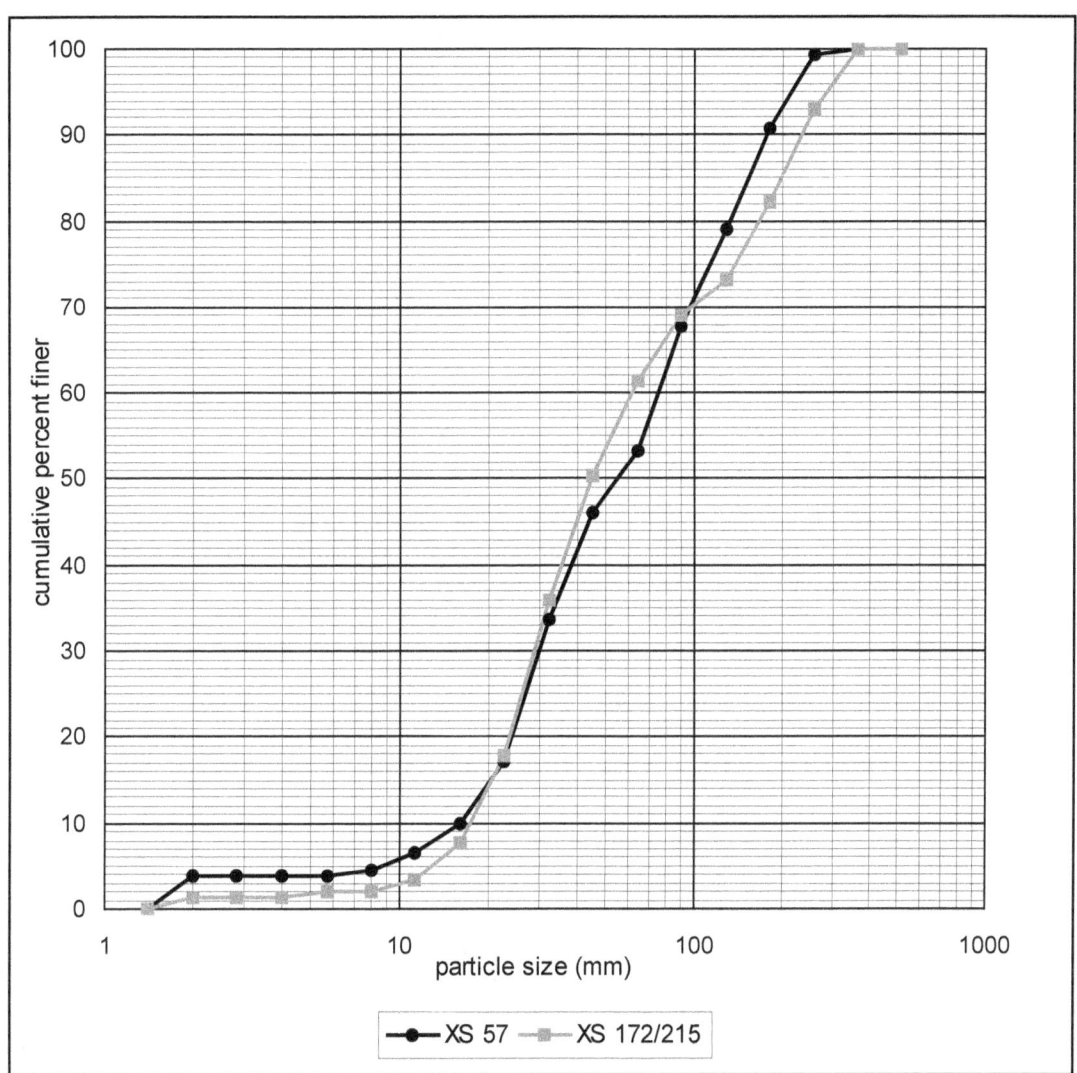

Figure H.6.   Bed Material Gradation.

Step 3. Check for Dynamic Equilibrium.

The qualitative assessment for dynamic equilibrium involves three components:

1. Watershed reconnaissance for changes in supply.

2. Project reach sediment transport assessment.

3. Field observations of the project reach.

The watershed reconnaissance is to identify changes in the watershed that may result in changes in sediment supply. As described in the site summary, activities in the watershed include logging, cattle grazing, and recreational use. Timber removed from the watershed will be replaced with new trees over time and the other uses are not reported to be intensive and are not forecast to change significantly.

The project reach sediment transport assessment examines potential changes in discharge, slope, and D_{50} throughout the project reach in the context of Lane's proportional relationship to sediment transport (Equation 7.2). Discharge throughout the reach is invariant and the sediment size appears relatively homogeneous based on the two samples summarized in Figure H.6. Average slopes in the project reach range from 0.006 ft/ft to 0.043 ft/ft (see Table H.3). The two lowest gradient sections (below cross-sections 399 and 342) and the two highest gradient sections (below cross-sections 307 and 215) may have been affected by the culvert itself. Slope variations are not considered to be an issue for sediment transport.

Field observations are to identify any indicators of instability. No instabilities were reported.

Taking the components together, there is no clear concern that the project reach is experiencing instability or disequilibrium in the sediment transport. The design for the passage culvert may improve sediment transport equilibrium.

Step 4. Analyze and Mitigate Channel Instability.

Based on the assessment in Step 3, this step is unnecessary.

Step 5. Align and Size Culvert for Q_p.

CFLHD criteria allows for a headwater depth to culvert rise ratio (HW/D ratio) of up to 1.5 for culverts with a rise less than or equal to 4 ft and up to 1.2 for culverts with a rise of greater than 4 ft. CDOT allows a 1.3 HW/D ratio for culverts between 3 and 5 ft, 1.2 for culverts between 5.5 and 7 ft, and 1.0 for culverts larger than 7 ft. The more stringent of these two criteria sets will be applied. An embedded CMP culvert will be designed with these criteria.

HY-8 is used to assess the existing culvert, which is a 3-ft diameter CMP culvert on a slope of 3.9 percent. For the peak design flow, Q_P, of 103 cfs the existing culvert has a HW/D ratio of 2.4 and the road overtops. Therefore, the existing culvert does not meet hydraulic criteria for the site.

The horizontal alignment of the existing culvert will be maintained.

The desired vertical alignment of the replacement culvert is established by evaluation of the vertical profile of the stream and the existing culvert profile. Figure H.5 shows the project reach longitudinal profile, which exhibits, on average, a project reach slope of 0.0267 ft/ft. For the initial design trial, the culvert will be laid on the average slope for the project reach (0.0267 ft/ft) rather than the steeper slope of the existing culvert. The existing culvert slope is undesirably steep. Placing the culvert on a slope more typical for the project reach will reduce outlet velocity and improve the potential for achieving bed stability in the culvert at Q_H (Step 6). This will require raising the outlet 0.28 ft and lowering the inlet 0.3 ft. During installation of the culvert, these adjustments will be made in conjunction with filling the scour hole at the outlet and removing deposits at the inlet caused by the existing culvert.

An initial CMP culvert diameter of 6.5 ft is estimated considering that the existing 3-ft diameter CMP is inadequate and a 2 ft minimum embedment is required. The embedment criteria for a circular culvert are 30 percent of the culvert rise giving an embedded depth of 0.3 x 6.5 ft = 1.95 ft. However, the minimum embedment depth 2.0 ft will be used for this design. Inlet and outlet elevations for the existing and replacement culvert are summarized in Table H.5.

Table H.5.   Inlet and Outlet Elevations for Existing and Replacement Culverts.

Description	Inlet	Outlet
Existing Culvert Invert	94.48	92.67
Replacement Culvert Bed	94.18	92.95
Replacement Culvert Invert	92.18	90.95

A bed gradation must be selected. Since we have two similar gradation samples from the site, an intermediate design gradation will be selected. The slope of the creek at XS 57 is closer to the proposed culvert bed slope than the steeper area between XS 172 and 215 so this gradation is a good starting point. To be modestly conservative, the average of the two gradations is selected. See Table H.6.

To control bed interstitial flow, it is recommended that the D_5 fraction be no larger than 2 mm (sand, silt, and clay). Both existing gradations contain at least 1 percent of this fraction; an additional 4 percent must be added. The final bed gradation design is shown in the last column of Table H.6.

Table H.6. Bed Gradation Design.

Quantile	XS 57 (mm)	XS 172/215 (mm)	Design (mm)	Design with added fines (mm)	Design with added fines (ft)
D_{95}	218	285	250	250	0.82
D_{84}	151	194	170	170	0.56
D_{50}	55	45	50	50	0.16
D_{16}	22	21	21	20	0.066
D_5	13	13	13	2	0.0066

The culvert embedment must be no less than 2 times the D_{95}. The D_{95} from Table H.6 is 250 mm (0.82 ft). Since 2 times D_{95} is less than 2.0 ft, the embedment depth is satisfactory.

A Manning's n is needed to estimate the roughness of the bed material in the culvert. The Limerinos equation (C.1) is used over a range of depths using the depth above the bed in place of the hydraulic radius as recommended in Appendix C. Since we do not know the flow depth, the roughness value will be calculated over a range of depths. The calculation for a depth of 1 ft is as follows. Table H.7 summarizes Manning's n for a range of depths.

$$n = \frac{\alpha \, y^{1/6}}{1.16 + 2 \log\left(\dfrac{y}{D_{84}}\right)} = \frac{0.0926(1.0)^{1/6}}{1.16 + 2 \log\left(\dfrac{1.0}{0.56}\right)} = 0.056$$

Table H.7. Manning's n for Bed Material (D_{84} = 0.56 ft).

Depth (ft)	Manning's n
1.0	0.056
1.5	0.049
2.0	0.046
2.5	0.044
3.0	0.042
3.5	0.041
4.0	0.041

HY-8 is used to analyze the culvert. The Manning's n corresponding with normal depth in the culvert at the design flow is used. For a 6.5 ft CMP culvert with 2.0 ft of embedment the normal depth at Q_P is 2.25 ft, therefore, a Manning's n of 0.045 (Table H.7) is appropriate for the bed. For these conditions the culvert operates in outlet control with a headwater depth of 3.47 ft. The HW/D ratio is 3.47/(6.5-2.0) = 0.77, which is less than or equal to the 1.2 maximum criteria. (As will be evident later in the example, a smaller culvert with a higher HW/D ratio will fail subsequent tests so a smaller size is not attempted at this time.) The estimated headwater also does not overtop the road or result in a redirection of flows away from the culvert. This culvert has a higher rise than the existing, but minimum cover still appears to be available. Proceed to Step 6.

Step 6. Check Culvert Bed Stability at Q$_H$.

Since the proposed culvert bed is less than 3 percent, the modified permissible shear stress approach will be applied. The shear stress in the culvert is compared with the permissible shear stress for the bed material. To compute the permissible shear stress, the value for Shield's parameter must first be determined based on the shear velocity, V_*, Reynolds number, R_e, and Table 7.1. Because shear velocity is a function of depth, the normal depth in the culvert is iteratively determined along with the Manning's n.

The iteration is accomplished by assuming a normal depth, estimating the Manning's n based on that depth, and then applying HY-8 with that roughness to determine the normal depth. If the calculated depth matches the assumed depth, the iteration is completed. If not, a new depth is assumed and the process repeated. At Q$_H$ of 8.8 ft^3/s, the Manning's n is 0.06 and normal depth in the culvert equals 0.55 ft according to HY-8. Using Equations 7.4 and 7.5:

$$V_* = \sqrt{gyS} = \sqrt{32.2(0.55)(0.0267)} = 0.687 \text{ ft/s}$$

$$R_e = \frac{V_* D_{50}}{\nu} = \frac{0.687(0.16)}{1.217 \times 10^{-5}} = 9.03 \times 10^3$$

Consulting Table 7.1 with this Reynolds number results in $F_* = 0.047$.

The permissible shear stress is determined by Equation 7.6:

$$\tau_p = F_*(\gamma_s - \gamma) D_{84}^{0.3} D_{50}^{0.7} = 0.047(156 - 62.4)(0.56)^{0.3}(0.16)^{0.7} = 1.0 \text{ lb/ft}^2$$

Applied shear stress is estimated at the inlet and outlet of the culvert based on the estimated depths. HY-8 reports that the culvert is operating in outlet control with flow type 3-M2t at Q$_H$ (see section H.3.2). For this flow type, the inlet and outlet depths are considered to determine limiting shear conditions within the culvert. The inlet and outlet depths are taken from the water surface profile data for Q$_H$ in HY-8 and are 0.55 ft and 0.46 ft, respectively.

The energy slope, S, is estimated from Manning's equation once velocity and composite n are computed. The wetted perimeter of the wall is based on the circular arc.

For the inlet, the calculations are:

y = 0.55 ft

A = 3.405 ft^2

P_{bed} = 6.000 ft

P_{wall} = 1.155 ft

P_{total} = 7.155 ft

R_h = 3.4059/7.155 = 0.476 ft

$$n_{comp} = \left[\frac{P_{bed} n_{bed}^{1.5} + P_{wall} n_{wall}^{1.5}}{P_{bed} + P_{wall}} \right]^{2/3} = \left[\frac{6.000(0.06)^{1.5} + 1.155(0.024)^{1.5}}{7.155} \right]^{2/3} = 0.055$$

V = Q/A = 8.8/3.405 = 2.58 ft/s

$$S = \left(\frac{V(n)}{1.49\left(R_h^{2/3}\right)}\right)^2 = \left(\frac{2.58(0.055)}{1.49(0.476)^{2/3}}\right)^2 = 0.0245 \text{ ft / ft}$$

For the M2 profile, the inlet depth should be approaching normal depth where the culvert bed slope and energy slope should be the same at 0.0267 ft/ft. The above result confirms that we are at or near normal depth at the inlet.

The applied shear stress is computed using Equation 7.9:

$$\tau_d = \gamma y S = 62.4(0.55)(0.0245) = 0.9 \text{ lb / ft}^2$$

For the outlet, the calculations are:

y = 0.46 ft

A = 2.836 ft^2

P_{wall} = 0.970 ft

P_{bed} = 6.000 ft

P_{total} = 6.970 ft

R_h = 2.836/6.970 = 0.407 ft

$$n_{comp} = \left[\frac{P_{bed}n_{bed}^{1.5} + P_{wall}n_{wall}^{1.5}}{P_{bed} + P_{wall}}\right]^{2/3} = \left[\frac{6.000(0.06)^{1.5} + 0.970(0.024)^{1.5}}{6.970}\right]^{2/3} = 0.056$$

V = Q/A = 8.8/2.836 = 3.10 ft/s

$$S = \left(\frac{V(n)}{1.49\left(R_h^{2/3}\right)}\right)^2 = \left(\frac{3.10(0.056)}{1.49(0.407)^{2/3}}\right)^2 = 0.0451 \text{ ft / ft}$$

$$\tau_d = \gamma y S = 62.4(0.46)(0.0451) = 1.3 \text{ lb / ft}^2$$

The results are summarized in Table H.8. Comparing inlet and the outlet conditions, the highest shear stress is at the outlet estimated as 1.3 lb/ft^2. Since this is more than the permissible shear stress of 1.0 lb/ft^2, the culvert bed is not stable at Q_H.

Alternatively, if HEC-RAS is used to analyze the culvert, the culvert is reported to be operating under outlet control and the depths at the culvert inlet and outlet are computed to be 0.53 and 0.49 ft, respectively. (See section H.3.3.) Comparing inlet and the outlet conditions, the highest shear stress is at the outlet estimated as 1.1 lb/ft^2. (See Table H.8.) Since this is more than the permissible shear stress of 1.0 lb/ft^2, the culvert bed is not stable at Q_H.

Table H.8.   6.5 ft Culvert Inlet and Outlet Parameters at Q_H.

Parameter*	HY-8		HEC-RAS	
	Inlet	Outlet	Inlet	Outlet
y (ft)	0.55	0.46	0.53	0.49
V (ft/s)	2.58	3.10	2.68	2.91
S_e (ft/ft)	0.0245	0.0451	0.0276	0.0368
τ_d (lbs/ft^2)	0.8	1.3	0.9	1.1
τ_p (lbs/ft^2)	1.0	1.0	1.0	1.0

*Embedment=2.0 ft, S_o=0.0267 ft/ft, n_{bed}=0.060.

Regardless of which tool is applied, the culvert bed is determined not to be stable at Q_H and we must continue with Step 7.

Step 7. Check Channel Bed Mobility at Q_H.

The assessment in Step 6 concluded that the bed material in the culvert bottom is not stable at Q_H. In this step, we evaluate whether material is moving at this discharge in the upstream and the downstream channels. Table H.9 summarizes shear stress estimates at the cross-sections upstream and downstream using alternative methods: 1) a normal depth assumption and 2) HEC-RAS. The table also provides the computed shear stresses at the inlet and outlet of the culvert and the permissible shear stress for the bed material.

Table H.9.   Estimated Shear Stresses at Q_H.

Cross-section**	Normal Depth/HY-8 (lb/ft^2)	HEC-RAS (lb/ft^2)
567	1.3	1.1
472	1.1	1.6
399	0.8	0.7
342	1.8	2.5
307	1.1	0.8
Culvert Inlet*	0.8	0.9
Culvert Outlet*	1.3	1.1
215	1.2	2.0
172	0.7	0.6
125	1.0	0.9
57	1.1	1.1
τ_p (lbs/ft^2)	1.0	1.0

*6.5 ft CMP, embedment = 2.0 ft, n_{bed} = 0.060.
**n_{bed} = 0.060.

In accordance with the guidance for this step, we observe whether or not the shear stress in any channel cross-section is less than the permissible shear stress. The answer to this question is yes, regardless of which method is used, therefore we must return to Step 5 to redesign the culvert to achieve a stable bed.

Step 5. Align and Size Culvert for Q_p (Trial 2).

For this trial we will consider increasing the barrel diameter 1.0 ft to a 7.5-ft CMP. The embedment criteria for a circular culvert is 30 percent of the culvert rise giving an embedded depth of 0.3 x 7.5 ft = 2.25 ft using the design bed gradation in Table H.6. A Manning's n of 0.045 (from Trial 1) will be used for Q_P. A headwater check at Q_P is not needed because this test had already been satisfied with a smaller culvert barrel. Proceed to Step 6.

Step 6. Check Culvert Bed Stability at Q_H (Trial 2).

For Trial 2, we apply the same methodologies as we applied in Trial 1. At Q_H, the Manning's n remains 0.060 and the permissible shear remains 1.0 lb/ft^2.

Applied shear stress is estimated at the inlet and outlet of the culvert based on the estimated depths. The inlet and outlet depths are taken from the water surface profile data for Q_H in HY-8 and are 0.51 ft and 0.46 ft, respectively. The resulting shear stresses are summarized in Table H.10. The applied shear stress is less than or equal to the permissible shear stress, therefore the bed is stable.

Alternatively, if HEC-RAS is used to analyze the culvert, the depths at the culvert inlet and outlet are computed to be 0.49 and 0.49 ft, respectively. The resulting shear stresses are summarized in Table H.10. The applied shear stress is less than or equal to the permissible shear stress, therefore the bed is stable.

Table H.10.   7.5 ft Culvert Inlet and Outlet Parameters at Q_H.

Parameter*	HY-8		HEC-RAS	
	Inlet	Outlet	Inlet	Outlet
y (ft)	0.51	0.46	0.49	0.49
V (ft/s)	2.44	2.71	2.54	2.54
S_e (ft/ft)	0.0242	0.0338	0.0275	0.0275
τ_d (lbs/ft^2)	0.8	1.0	0.8	0.8
τ_p (lbs/ft^2)	1.0	1.0	1.0	1.0

*Embedment=2.25 ft, S_o=0.0267 ft/ft, n_{bed}=0.060.

Regardless of which tool is applied, the culvert bed is determined to be stable at Q_H and we proceed to Step 8.

Step 8. Check Culvert Bed Stability at Q_P.

The shear stress in the culvert is compared with the permissible shear stress for the bed material. To compute the permissible shear stress, the value for Shield's parameter must first be determined based on the shear velocity, V_*, Reynolds number, R_e, and Table 7.1. Because shear velocity is a function of depth, the normal depth in the culvert is iteratively determined along with the Manning's n. At Q_P, the Manning's n is 0.045 and normal depth in the culvert equals 2.1 ft. Using Equations 7.4, 7.5, and 7.6 as before yields a permissible shear stress for the bed material of 1.0 lbs/ft^2.

Applied shear stress is estimated at the inlet and outlet of the culvert based on the estimated depths. The inlet and outlet depths are taken from the water surface profile data for Q_H in HY-8 and are 2.04 ft and 1.84 ft, respectively. The resulting shear stresses are summarized in Table H.11. The applied shear stress is greater than the permissible shear stress, therefore the bed is not stable.

Alternatively, if HEC-RAS is used to analyze the culvert, the depths at the culvert inlet and outlet are computed to be 1.88 and 1.84 ft, respectively. The resulting shear stresses are summarized in Table H.11. The applied shear stress is greater than the permissible shear stress, therefore the bed is not stable.

Table H.11. 7.5 ft CMP Culvert Inlet and Outlet Parameters at Q_P.

Parameter*	HY-8		HEC-RAS	
	Inlet	Outlet	Inlet	Outlet
y (ft)	2.04	1.84	1.88	1.84
V (ft/s)	6.88	7.64	7.47	7.64
S_e (ft/ft)	0.0205	0.0276	0.0260	0.0276
τ_d (lbs/ft^2)	2.6	3.2	3.1	3.2
τ_p (lbs/ft^2)	1.0	1.0	1.0	1.0

*Embedment=2.25 ft, S_o=0.0267 ft/ft, n_{bed}=0.045.

Regardless of the method selected, the culvert bed is not stable at Q_P, therefore, we must continue with Step 9.

Step 9. Design Stable Bed for Q_P.

A stable bed design is attempted to resist the shear stresses at Q_P within the culvert. The bed will consist of a top layer of native material and an oversized underlayer. Design of the underlayer assumes the native top layer has been washed away at or before the peak of the hydrograph. It is assumed that natural replenishment cannot be relied on to restore the bed material in the culvert. If, however, site-specific analysis to the contrary is performed, the stable sublayer may be avoided.

As a first trial in designing the sublayer, select an oversized bed material that fits within the current culvert embedment of 2.25 ft. In accordance with the embedment criteria for Step 9, we would provide a 1 ft layer of native material leaving 1.25 ft for the oversized bed material. For a CMP culvert, the oversize layer minimum embedment is $1.5D_{95}$, therefore, D_{95} = 1.25 ft/1.5 = 0.83 ft. Using the relation in Equation 7.15c between D_{50} and D_{95} for an oversized bed, the D_{50} = D_{95}/1.9 = 0.83 ft/1.9 = 0.44 ft. However, we will learn that this bed will not be stable at Q_P.

As a second trial, select an oversized bed material that fits within an 8.5 ft CMP with a total embedment of 2.55 ft (30 percent of the culvert rise). In accordance with the embedment criteria, we would provide a 1 ft layer of native material leaving 1.55 ft for the oversized bed material. For a CMP culvert, the oversize layer minimum embedment is $1.5D_{95}$, therefore, D_{95} = 1.55 ft/1.5 = 1.0 ft. Using the relation in Equation 7.15c between D_{50} and D_{95} for an oversized bed, the D_{50} = D_{95}/1.9 = 1.0 ft/1.9 = 0.53 ft.

Assuming the native layer is washed out, we use HY-8 iteratively to determine that the normal depth in the culvert is 2.31 ft with a Manning's n of 0.057 for the oversize layer. From this, a Shield's parameter of 0.054 is determined. The permissible shear stress is calculated from Equation 7.16:

$$\tau_p = 1.1F_*(\gamma_s - \gamma) D_{50} = 1.1(0.054)(156\text{-}62.4)(0.53) = 2.9 \text{ lbs/ft}^2$$

Applied shear stress is estimated at the inlet and outlet of the culvert based on the estimated depths. The inlet and outlet depths are taken from the water surface profile data for Q_P in HY-8 and are 2.31 ft and 2.27 ft, respectively (see section H.3.4). The resulting shear stresses are summarized in Table H.12. The applied shear stress is less than the permissible shear stress, therefore the bed is stable.

Alternatively, if HEC-RAS is used to analyze the culvert, the depths at the culvert inlet and outlet are computed to be 2.13 and 2.68 ft, respectively (see section H.3.5). The resulting shear stresses are summarized in Table H.12. The applied shear stress is less than the permissible shear stress, therefore the bed is stable.

Table H.12. 8.5 ft CMP Culvert Inlet and Outlet Parameters at Q_P.

Parameter*	HY-8		HEC-RAS	
	Inlet	Outlet	Inlet	Outlet
y (ft)	2.31	2.27	2.13	2.68
V (ft/s)	5.73	5.84	6.26	4.88
S_e (ft/ft)	0.0160	0.0169	0.0215	0.0097
τ_d (lbs/ft^2)	2.3	2.4	2.86	1.6
τ_p (lbs/ft^2)	2.9	2.9	2.9	2.9

*Embedment=1.55 ft (native layer washed out), S_o=0.0267 ft/ft, n_{bed}=0.057.

Regardless of the method selected, the culvert bed is stable at Q_P, therefore, we continue by completing the oversized bed gradation design.

D_{84} is computed from Equation 7.15b:

$D_{84} = 1.4D_{50} = 1.4(0.53) = 0.74$ ft

The D_5 is taken to be no larger than 2 mm to limit interstitial flow. The D_{16} is selected to provide a transition between the D_{50} and D_5 sizes. A reasonable transition is determined graphically. Table H.13 summarizes the resulting gradation and compares it to the native bed gradation.

Table H.13. Oversize Stable Bed Design Gradation.

Quantile	Native (mm)	Oversize (mm)	Native (ft)	Oversize (ft)
D_{95}	250	305	0.82	1.00
D_{84}	170	226	0.56	0.74
D_{50}	50	162	0.16	0.53
D_{16}	20	36	0.066	0.12
D_5	2	2	0.0066	0.0066

Step 10. Check Culvert Velocity at Q_H.

A check is conducted to verify that the culvert velocity is less than or equal to at least part of the upstream or downstream channel. The check is satisfied if the culvert inlet and outlet velocities are within the range of the cross-section velocities. Our culvert embedment has both an upper native bed layer and a lower oversized layer. Since this check is performed at Q_H, it is assumed that the native bed material layer is present.

HY-8 reports that the culvert is in outlet control with a 3-M2t profile in the barrel (See section H.3.6) and the inlet and outlet velocities from the water surface profile data in HY-8 are shown in Table H.14. The velocity in the 46 ft culvert varies from 2.4 to 2.7 ft/s. Upstream of the culvert in the reaches indicated by cross-sections 342 and 307 there are higher velocities of 2.4 to 2.9 ft/s through a distance of 61 ft. Therefore, the velocity in the culvert does not present conditions more severe than are found elsewhere in the project reach.

HEC-RAS also reports outlet control with inlet and outlet velocities ranging from 2.3 to 2.4 ft/s (see section H.3.7) as shown in Table H.14. In addition to cross-section 342, cross-sections 567, 472, 215, and 57 all exhibit higher velocities than estimated in the culvert. Several of the reach lengths associated with each cross-section are also longer than the culvert. Therefore, the velocity check is satisfied.

For performing the velocity check, the HEC-RAS will generally be preferred because the channel velocities are based on a water surface profile analysis rather than the simplifying assumption of normal depth at each cross-section. In this example, however, both techniques result in the same conclusion. Since the check is satisfied, we proceed to Step 11.

Table H.14. Velocity Estimates at Q_H.

Cross-section**	Applicable Reach Length (ft)	Normal Depth/HY-8 (ft/s)	HEC-RAS (ft/s)
567	95	2.48	2.41
472	73	2.38	2.93
399	57	1.57	1.69
342	35	2.86	3.46
307	26	2.40	2.00
Culvert Inlet*	23	2.68	2.40
Culvert Outlet*	23	2.42	2.34
215	43	2.37	3.04
172	47	2.02	1.98
125	68	2.12	2.16
57	57	2.43	2.52

*8.5 ft CMP, embedment = 2.55 ft, S_o = 0.0267 ft/ft, n_{bed} =0.060.
**n_{bed} =0.060

Step 11. Check Culvert Water Depth at Q_L.

A check is conducted to verify that the culvert depth is greater than or equal to at least part of the upstream or downstream channel. Table H.15 summarizes the maximum depths estimated at each cross-section and within the culvert by both methods. According to the normal depth/HY-8 methods, the depths in the culvert bed are shallower than those in the upstream and downstream channel meaning that the culvert is the limiting location in terms of depth and the depth check is not satisfied. However, according to the HEC-RAS analysis, cross-section 215 exhibits the lowest depth meaning the culvert is not the limiting location and the depth check is satisfied according to HEC-RAS.

As with the velocity check, HEC-RAS is generally preferred because the channel depths are based on a water surface profile analysis rather than the simplifying assumption of normal depth at each cross-section. The conclusion regarding the depth check hinges on the two methods assessment of the depth at cross-section 215 and at the culvert inlet. The differences at cross-section 215 derive from a normal depth versus water surface profile methods. The difference at the culvert inlet results from alternative computations of normal depth, which both programs estimate should exist at the culvert inlet. Given the mixed results, we will proceed with the conclusion that the check for water depth is not satisfied. Proceed to Step 12 to provide a low-flow channel.

Table H.15. Maximum Depth Estimates at Q_L.

Cross-section**	Normal Depth/HY-8 (ft)	HEC-RAS (ft)
567	0.30	0.34
472	0.30	0.23
399	0.58	0.63
342	0.28	0.24
307	0.24	0.28
Culvert Inlet*	0.09	0.13
Culvert Outlet*	0.17	0.16
215	0.18	0.13
172	0.29	0.31
125	0.30	0.30
57	0.25	0.25

*8.5 ft CMP, embedment=2.55 ft, S_o=0.0267 ft/ft, n_{bed} = 0.060.
*n_{bed} = 0.060.

Step 12. Provide Low-flow Channel in Culvert.

To increase the depth in the culvert bed, add a triangular low-flow channel with side slopes of 1:8 (V:H). This will provide a thalweg 0.5 ft deeper in the center of the culvert. However, noting that the D_{84} of the native bed material is 0.55 ft the "construction" of such a small channel will require careful manual work. The stream will likely form and maintain a low flow channel over time.

Step 13. Review Design.

A 8.5-ft (2.60-m) CMP with a 2.55-ft (0.78-m) embedment on a 2.67 percent slope is proposed to replace the 3.0-ft (0.91-m) CMP culvert on a 3.9 percent slope. The embedment is characterized by a 1 ft (0.3 m) thick layer of native bed material with an oversized under layer 1.55 ft (0.48 m) in thickness to provide stability at Q_P. A low-flow channel is to be created to maintain depths in the culvert at Q_L.

Modification of the inlet and outlet areas to fill the scour hole and remove excess sediment is needed to properly place the new culvert. No change to the road profile is needed.

Alternative culvert shapes and materials may also be considered. A concrete box or pipe arch may offer an option to maintain a sufficiently wide span to meet the stability, velocity, and depth criteria with a lower rise.

H.3 SUPPORTING DOCUMENTATION

H.3.1 Surveyed Cross-Sections.

H.3.2 HY-8 Report for 6.5 ft CMP at Q_H.

H.3.3 HEC-RAS Output for 6.5 ft CMP at Q_H.

H.3.4 HY-8 Report for 8.5 ft CMP at Q_P.

H.3.5 HEC-RAS Output for 8.5 ft CMP at Q_P.

H.3.6 HY-8 Report for 8.5 ft CMP with Oversized Bed at Q_H.

H.3.7 HEC-RAS Output for 8.5 ft CMP with Oversized Bed at Q_H.

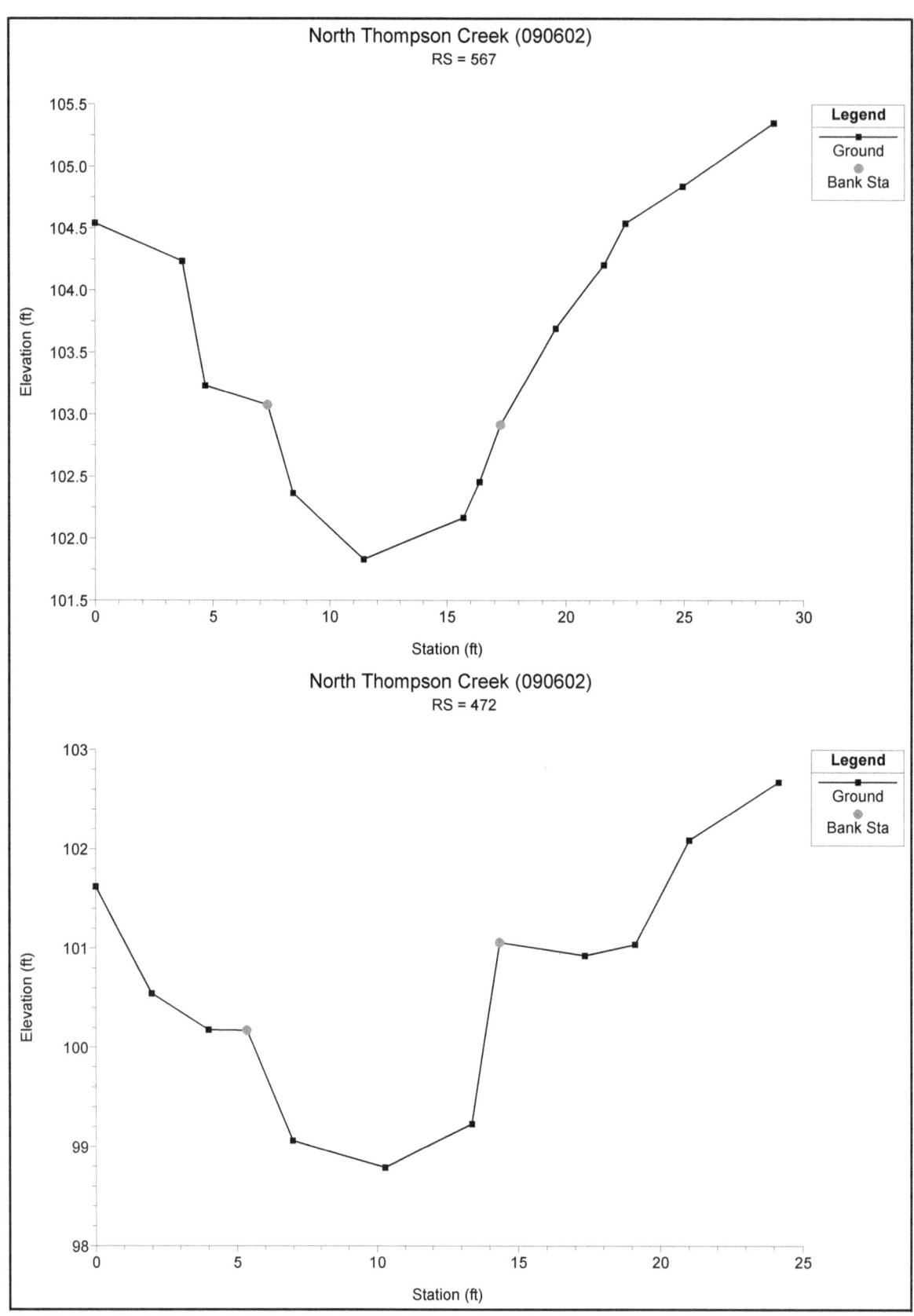

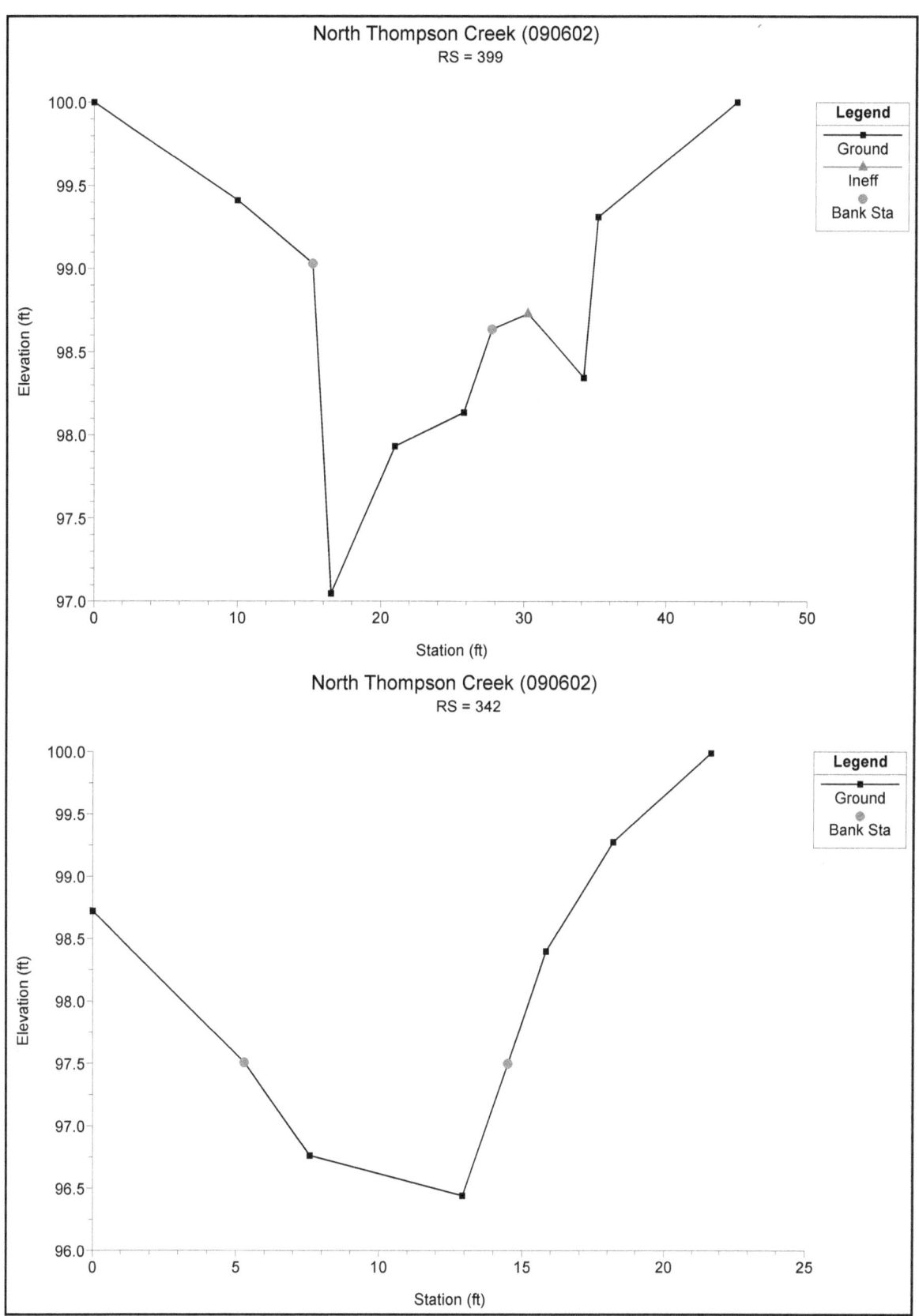

H-19

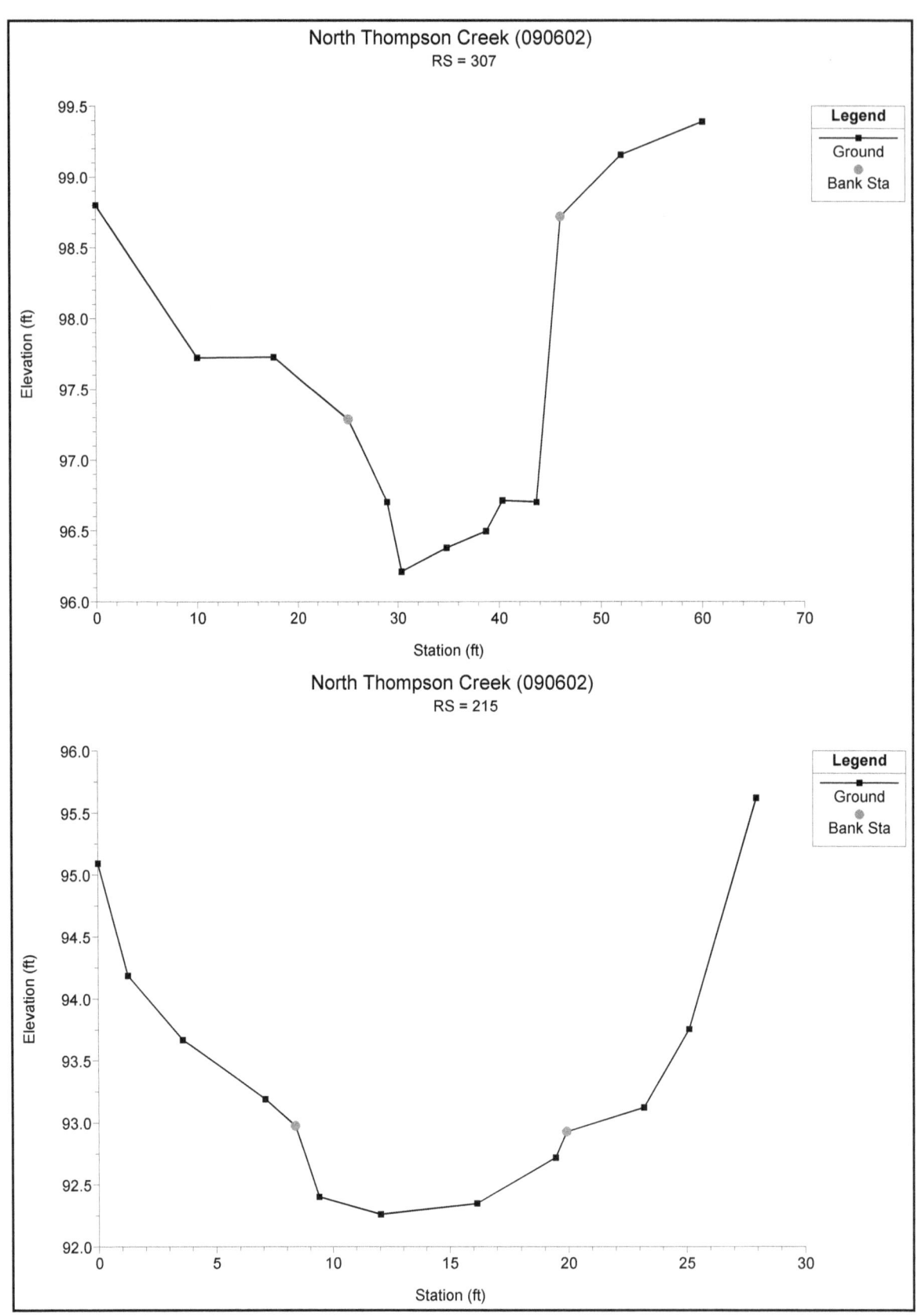

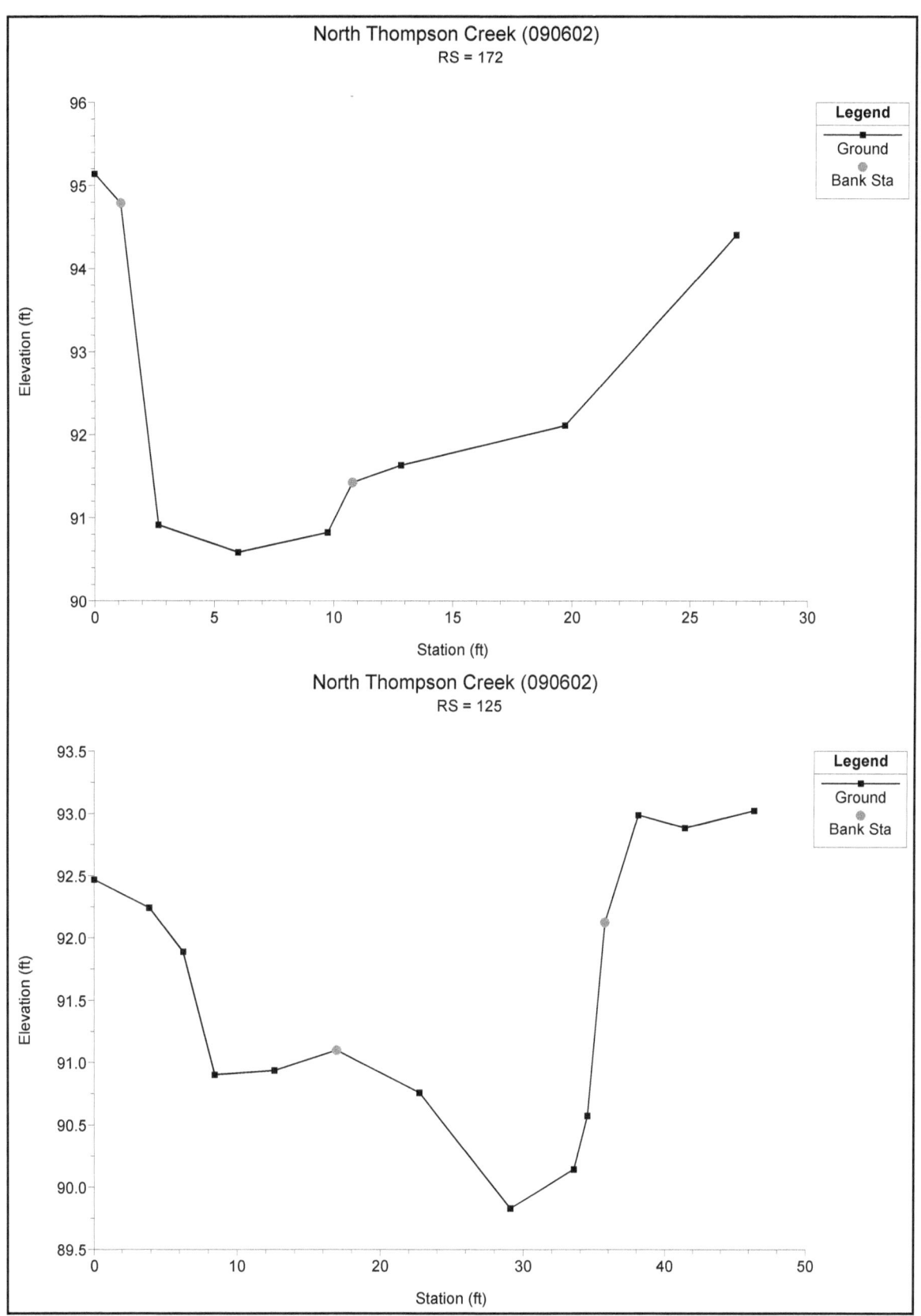

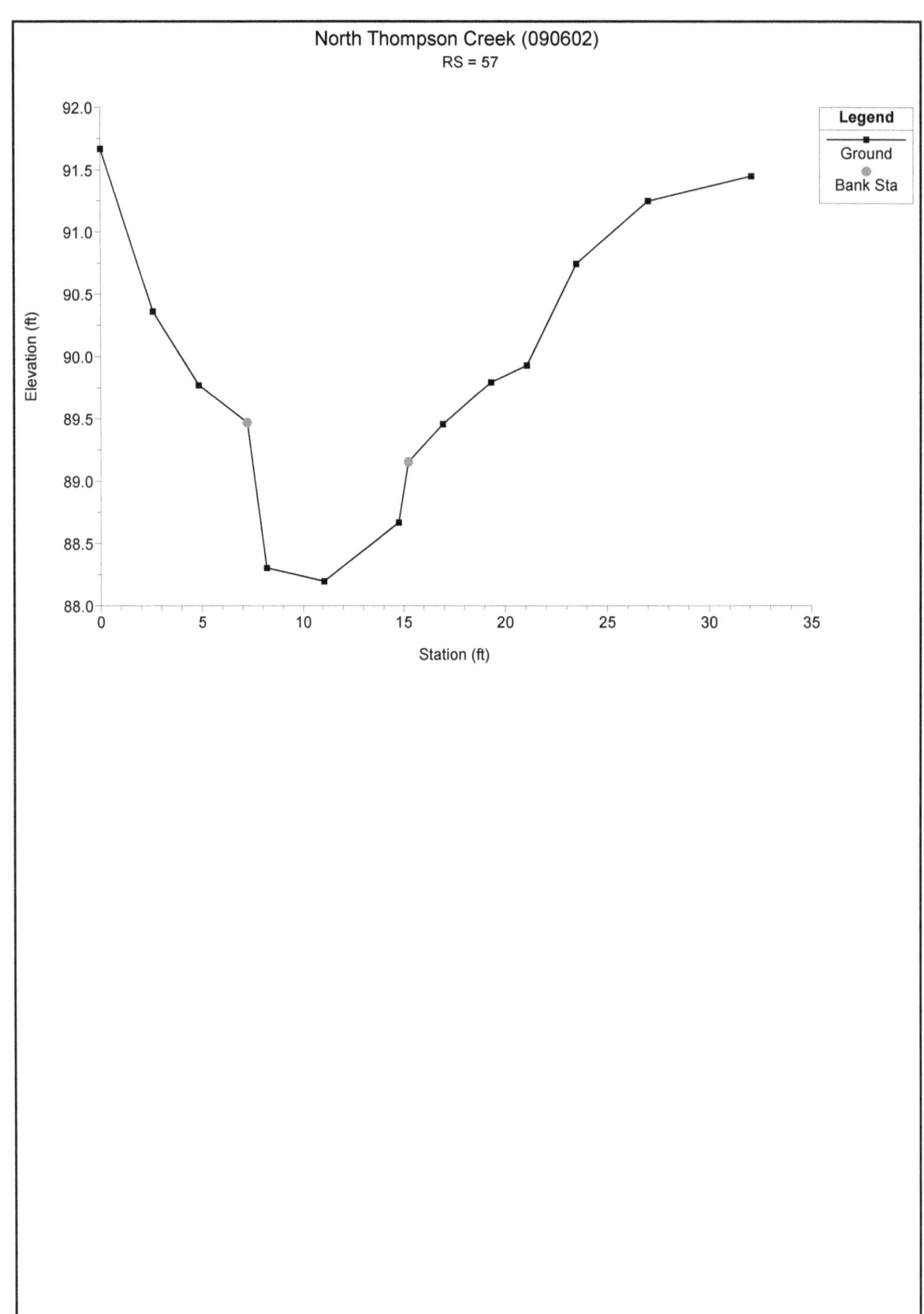

North Thompson Creek (090602)
RS = 57

Table 1 - Culvert Summary Table: 6.5' CMP emb=2.0' n=0.060

Total Discharge (cfs)	Culvert Discharge (cfs)	Headwater Elevation (ft)	Inlet Control Depth (ft)	Outlet Control Depth (ft)	Flow Type	Normal Depth (ft)	Critical Depth (ft)	Outlet Depth (ft)	Tailwater Depth (ft)	Outlet Velocity (ft/s)	Tailwater Velocity (ft/s)
0.00	0.00	94.18	0.000	-1.230	0-NF	0.000	0.000	0.000	0.000	0.000	0.000
0.88	0.88	94.30	0.062	0.115	3-M1t	0.065	0.039	0.162	0.162	0.921	0.859
1.76	1.76	94.37	0.122	0.187	3-M1t	0.131	0.077	0.214	0.214	1.380	1.116
2.64	2.64	94.43	0.179	0.254	3-M1t	0.196	0.116	0.256	0.256	1.722	1.296
3.52	3.52	94.50	0.235	0.320	3-M1t	0.262	0.154	0.294	0.294	1.987	1.421
4.40	4.40	94.58	0.289	0.397	3-M1t	0.327	0.193	0.329	0.329	2.212	1.515
5.28	5.28	94.65	0.343	0.471	3-M2t	0.392	0.231	0.361	0.361	2.417	1.594
6.16	6.16	94.79	0.397	0.609	3-M2t	0.454	0.270	0.389	0.389	2.607	1.664
7.04	7.04	94.81	0.452	0.631	3-M2t	0.486	0.308	0.416	0.416	2.785	1.727
7.92	7.92	94.85	0.511	0.673	3-M2t	0.518	0.347	0.440	0.440	2.953	1.784
8.80	8.80	94.90	0.576	0.718	3-M2t	0.550	0.385	0.464	0.464	3.113	1.836

```
************************************************************************
```
Inlet Elevation (invert): 94.18 ft,   Outlet Elevation (invert): 92.95 ft

Culvert Length: 46.02 ft,   Culvert Slope: 0.0267
```
************************************************************************
```

Site Data - 6.5' CMP emb=2.0' n=0.060

Site Data Option: Culvert Invert Data

Inlet Station: 0.00 ft

Inlet Elevation: 92.18 ft

Outlet Station: 46.00 ft

Outlet Elevation: 90.95 ft

Number of Barrels: 1

Culvert Data Summary - 6.5' CMP emb=2.0' n=0.060

Barrel Shape: Circular

Barrel Diameter: 6.50 ft

Barrel Material: Corrugated Steel

Embedment: 24.00 in

Barrel Manning's n: 0.0240 (top and sides)

Manning's n: 0.0600 (bottom)

Inlet Type: Conventional

Inlet Edge Condition: Square Edge with Headwall

Inlet Depression: None

Table 2 - Downstream Channel Rating Curve (Crossing: N Thompson, n=0.06)

Flow (cfs)	Water Surface Elev (ft)	Depth (ft)	Velocity (ft/s)	Shear (psf)	Froude Number
0.00	92.95	0.00	0.00	0.00	0.00
0.88	93.11	0.16	0.86	0.27	0.48
1.76	93.16	0.21	1.12	0.36	0.51
2.64	93.21	0.26	1.30	0.42	0.53
3.52	93.24	0.29	1.42	0.49	0.55
4.40	93.28	0.33	1.51	0.55	0.56
5.28	93.31	0.36	1.59	0.60	0.56
6.16	93.34	0.39	1.66	0.65	0.57
7.04	93.37	0.42	1.73	0.69	0.57
7.92	93.39	0.44	1.78	0.73	0.58
8.80	93.41	0.46	1.84	0.77	0.58

Tailwater Channel Data - N Thompson, n=0.06

Tailwater Channel Option: Irregular Channel

Roadway Data for Crossing: N Thompson, n=0.06

Roadway Profile Shape: Constant Roadway Elevation

Crest Length: 99.00 ft

Crest Elevation: 101.66 ft

Roadway Surface: Paved

Roadway Top Width: 23.00 ft

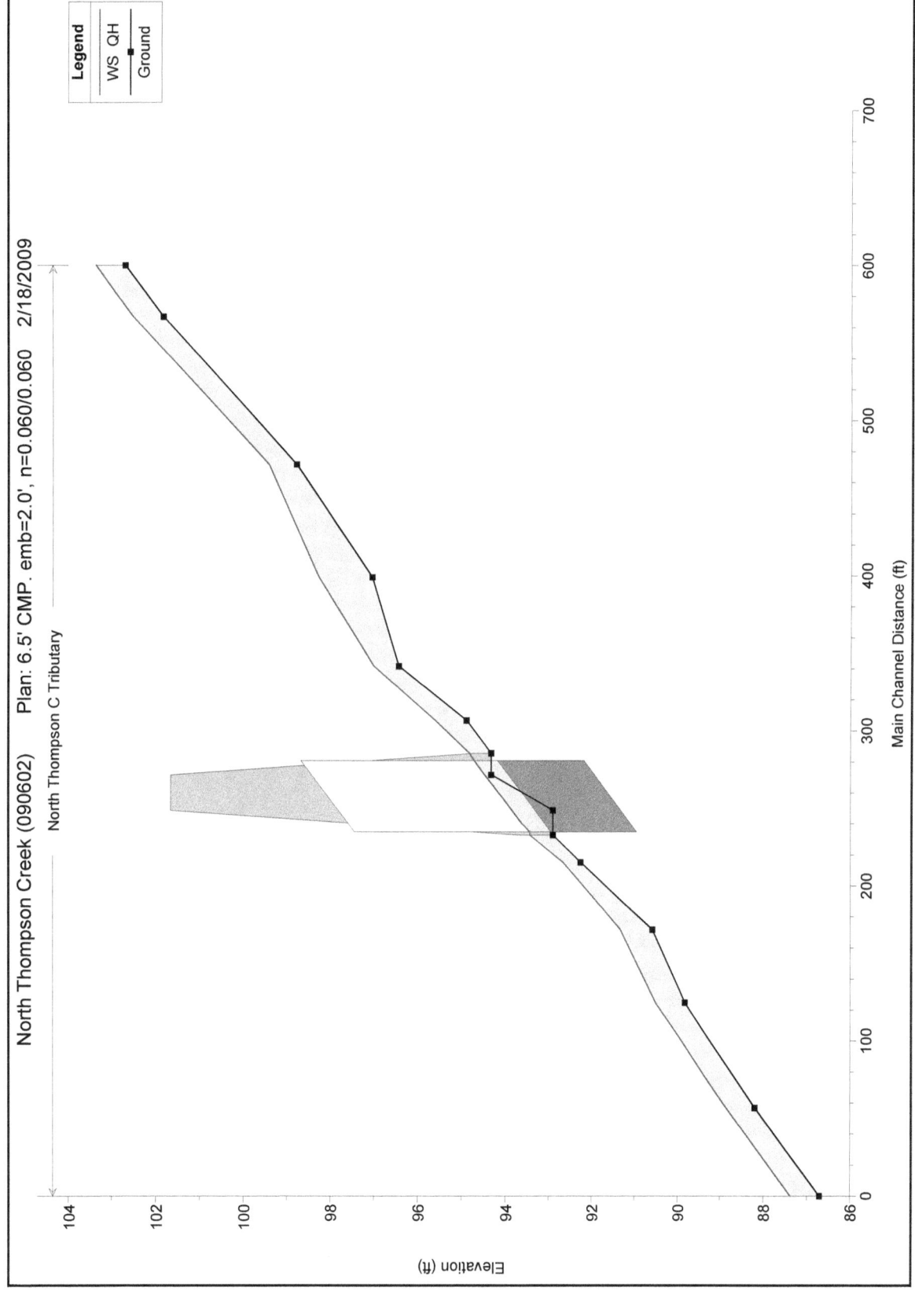

H.3.3. HEC-RAS Output for 6.5 ft Culvert at QH.

H.3.3. HEC-RAS Output for 6.5 ft Culvert at QH.

HEC-RAS Plan: 6.5' 2.0' 0.060 River: North Thompson C Reach: Tributary Profile: QH

Reach	River Sta	Profile	Q Total (cfs)	Min Ch El (ft)	W.S. Elev (ft)	Max Chl Dpth (ft)	Hydr Depth (ft)	Mann Wtd Chnl	Crit W.S. (ft)	E.G. Elev (ft)	E.G. Slope (ft/ft)	Vel Chnl (ft/s)	Flow Area (sq ft)	Top Width (ft)	Froude # Chl
Tributary	600	QH	8.80	102.69	103.37	0.68	0.44	0.057	103.25	103.46	0.026580	2.43	3.63	8.31	0.65
Tributary	567	QH	8.80	101.83	102.52	0.69	0.44	0.056		102.61	0.025444	2.41	3.65	8.32	0.64
Tributary	472	QH	8.80	98.79	99.42	0.63	0.43	0.057	99.35	99.55	0.041434	2.93	3.01	7.02	0.79
Tributary	399	QH	8.80	97.05	98.27	1.23	0.49	0.049	97.97	98.32	0.008911	1.69	5.20	10.63	0.43
Tributary	342	QH	8.80	96.44	97.01	0.57	0.36	0.057	97.01	97.20	0.071477	3.46	2.54	6.97	1.01
Tributary	307	QH	8.80	94.88	95.60	0.72	0.50	0.060	95.38	95.66	0.016938	2.00	4.39	8.76	0.50
Tributary	286	QH	8.80	94.32	94.82	0.50	0.33	0.060	94.82	94.98	0.077646	3.29	2.67	8.00	1.00
Tributary	260		Culvert												
Tributary	233	QH	8.80	92.90	93.44	0.54	0.47	0.056	93.32	93.53	0.024373	2.50	3.52	16.30	0.64
Tributary	215	QH	8.80	92.26	92.67	0.41	0.29	0.060	92.67	92.81	0.080273	3.04	2.90	10.09	1.00
Tributary	172	QH	8.80	90.59	91.32	0.73	0.55	0.056	91.09	91.38	0.013199	1.98	4.45	8.10	0.47
Tributary	125	QH	8.80	89.83	90.51	0.68	0.41	0.056		90.58	0.022114	2.16	4.08	9.94	0.59
Tributary	57	QH	8.80	88.20	88.86	0.66	0.48	0.057	88.73	88.96	0.026202	2.52	3.49	7.23	0.64
Tributary	0	QH	8.80	86.71	87.38	0.67	0.48	0.057	87.25	87.48	0.026048	2.52	3.49	7.23	0.64

H-28

H.3.3. HEC-RAS Output for 6.5 ft Culvert at QH.

Plan: 6.5' 2.0' 0.060   North Thompson C   Tributary RS: 260   Culv Group: Culvert #1   Profile: QH

Q Culv Group (cfs)	8.80	Culv Full Len (ft)	
# Barrels	1	Culv Vel US (ft/s)	2.67
Q Barrel (cfs)	8.80	Culv Vel DS (ft/s)	2.92
E.G. US. (ft)	94.88	Culv Inv El Up (ft)	92.18
W.S. US. (ft)	94.82	Culv Inv El Dn (ft)	90.95
E.G. DS (ft)	93.53	Culv Frctn Ls (ft)	1.25
W.S. DS (ft)	93.44	Culv Exit Loss (ft)	0.04
Delta EG (ft)	1.34	Culv Entr Loss (ft)	0.06
Delta WS (ft)	1.38	Q Weir (cfs)	
E.G. IC (ft)	94.72	Weir Sta Lft (ft)	
E.G. OC (ft)	94.88	Weir Sta Rgt (ft)	
Culvert Control	Outlet	Weir Submerg	
Culv WS Inlet (ft)	94.71	Weir Max Depth (ft)	
Culv WS Outlet (ft)	93.44	Weir Avg Depth (ft)	
Culv Nml Depth (ft)	2.53	Weir Flow Area (sq ft)	
Culv Crt Depth (ft)	2.40	Min El Weir Flow (ft)	101.67

Errors Warnings and Notes

Warning:	During subcritical analysis, the water surface upstream of culvert went to critical depth.

Table 1 - Culvert Summary Table: 8.5' CMP emb=1.55' n=0.057

Total Discharge (cfs)	Culvert Discharge (cfs)	Headwater Elevation (ft)	Inlet Control Depth (ft)	Outlet Control Depth (ft)	Flow Type	Normal Depth (ft)	Critical Depth (ft)	Outlet Depth (ft)	Tailwater Depth (ft)	Outlet Velocity (ft/s)	Tailwater Velocity (ft/s)
0.00	0.00	93.18	0.000	-0.230	0-NF	0.000	0.000	0.000	0.000	0.000	0.000
10.30	10.30	93.84	0.467	0.657	3-M1t	0.486	0.320	1.435	0.435	0.966	2.362
20.60	20.60	94.30	0.920	1.122	3-M1t	0.827	0.640	1.596	0.596	1.721	2.964
30.90	30.90	94.63	1.257	1.447	3-M1t	1.058	0.831	1.721	0.721	2.379	3.382
41.20	41.20	94.91	1.508	1.735	3-M1t	1.289	0.995	1.827	0.827	2.972	3.705
51.50	51.50	95.17	1.737	1.995	3-M1t	1.487	1.159	1.920	0.920	3.517	3.971
61.80	61.80	95.42	1.952	2.236	3-M1t	1.660	1.323	2.004	1.004	4.026	4.198
72.10	72.10	95.64	2.152	2.463	3-M1t	1.833	1.461	2.079	1.079	4.513	4.416
82.40	82.40	95.86	2.339	2.679	3-M1t	2.005	1.581	2.148	1.148	4.978	4.619
92.70	92.70	96.06	2.518	2.883	3-M1t	2.163	1.701	2.212	1.212	5.423	4.807
103.00	103.00	96.26	2.691	3.081	3-M2t	2.308	1.821	2.274	1.274	5.848	4.980

****

Inlet Elevation (invert): 93.18 ft,   Outlet Elevation (invert): 91.95 ft

Culvert Length: 46.02 ft,   Culvert Slope: 0.0267

****

Site Data - 8.5' CMP emb=1.55' n=0.057

Site Data Option: Culvert Invert Data

Inlet Station: 0.00 ft

Inlet Elevation: 91.63 ft

Outlet Station: 46.00 ft

Outlet Elevation: 90.40 ft

Number of Barrels: 1

Culvert Data Summary - 8.5' CMP emb=1.55' n=0.057

Barrel Shape: Circular

Barrel Diameter: 8.50 ft

Barrel Material: Corrugated Steel

Embedment: 18.60 in

Barrel Manning's n: 0.0240 (top and sides)

Manning's n: 0.0570 (bottom)

Inlet Type: Conventional

Inlet Edge Condition: Square Edge with Headwall

Inlet Depression: None

Table 2 - Downstream Channel Rating Curve (Crossing: N Thompson, n=0.045)

Flow (cfs)	Water Surface Elev (ft)	Depth (ft)	Velocity (ft/s)	Shear (psf)	Froude Number
0.00	92.95	0.00	0.00	0.00	0.00
10.30	93.39	0.44	2.36	0.72	0.77
20.60	93.55	0.60	2.96	0.99	0.81
30.90	93.67	0.72	3.38	1.20	0.84
41.20	93.78	0.83	3.70	1.37	0.86
51.50	93.87	0.92	3.97	1.53	0.88
61.80	93.95	1.00	4.20	1.67	0.89
72.10	94.03	1.08	4.42	1.79	0.90
82.40	94.10	1.15	4.62	1.91	0.91
92.70	94.16	1.21	4.81	2.01	0.92
103.00	94.22	1.27	4.98	2.11	0.92

Tailwater Channel Data - N Thompson, n=0.045

Tailwater Channel Option: Irregular Channel

Roadway Data for Crossing: N Thompson, n=0.045

Roadway Profile Shape: Constant Roadway Elevation

Crest Length: 99.00 ft

Crest Elevation: 101.66 ft

Roadway Surface: Paved

Roadway Top Width: 23.00 ft

H.3.5. HEC-RAS Output for 8.5 ft CMP at QP.

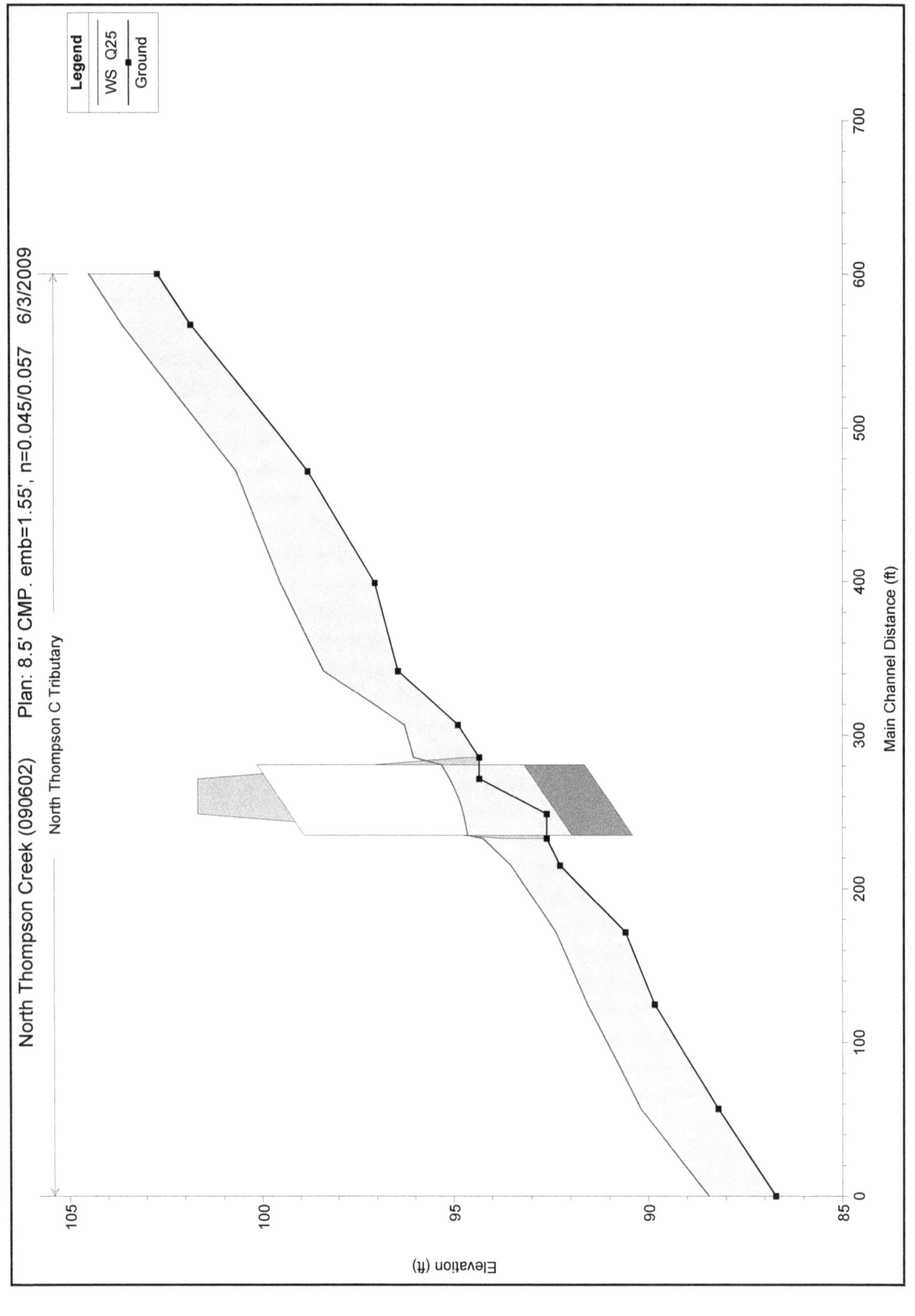

North Thompson Creek (090602)   Plan: 8.5' CMP. emb=1.55', n=0.045/0.057   6/3/2009

North Thompson C Tributary

Legend

WS Q25
Ground

Main Channel Distance (ft)

Elevation (ft)

H.3.5. HEC-RAS Output for 8.5 ft CMP at QP.

HEC-RAS Plan: 8.5 1.55 0.045 River: North Thompson C Reach: Tributary Profile: Q25

Reach	River Sta	Profile	Q Total (cfs)	Min Ch El (ft)	W.S. Elev (ft)	Max Chl Dpth (ft)	Hydr Depth (ft)	Mann Wtd Chnl	Crit W.S. (ft)	E.G. Elev (ft)	E.G. Slope (ft/ft)	Vel Chnl (ft/s)	Flow Area (sq ft)	Top Width (ft)	Froude # Chl
Tributary	600	Q25	103.00	102.69	104.48	1.79	1.08	0.043	104.55	105.19	0.026050	6.91	16.25	15.05	1.02
Tributary	567	Q25	103.00	101.83	103.61	1.78	1.08	0.042	103.69	104.33	0.025839	6.94	16.17	15.04	1.02
Tributary	472	Q25	103.00	98.79	100.66	1.87	1.16	0.042	100.81	101.55	0.032843	7.68	14.31	12.36	1.11
Tributary	399	Q25	103.00	97.05	99.52	2.47	0.97	0.040	99.15	99.76	0.007569	4.19	28.84	29.88	0.58
Tributary	342	Q25	103.00	96.44	98.38	1.94	1.18	0.040	98.38	99.05	0.019246	6.74	16.82	14.30	0.94
Tributary	307	Q25	103.00	94.88	96.27	1.39	0.92	0.060	96.65	97.52	0.153700	8.98	11.47	12.46	1.65
Tributary	286	Q25	103.00	94.32	96.04	1.72	1.09	0.055	96.04	96.58	0.043965	5.93	17.37	15.96	1.00
Tributary	260	Culvert													
Tributary	233	Q25	103.00	92.60	94.23	1.63	1.53	0.046	94.23	95.00	0.026903	7.03	14.66	24.79	1.00
Tributary	215	Q25	103.00	92.26	93.52	1.26	0.78	0.050	93.74	94.32	0.056874	7.44	15.37	19.71	1.26
Tributary	172	Q25	103.00	90.59	92.36	1.78	1.00	0.043	92.41	92.92	0.022966	6.40	18.49	18.40	0.91
Tributary	125	Q25	103.00	89.83	91.56	1.73	0.91	0.043	91.41	91.85	0.015810	4.53	25.73	28.30	0.76
Tributary	57	Q25	103.00	88.20	90.20	2.00	1.04	0.040	90.20	90.75	0.015743	6.35	19.33	18.64	0.85
Tributary	0	Q25	103.00	86.71	88.44	1.73	0.87	0.043	88.71	89.39	0.037670	8.10	14.43	16.65	1.18

H.3.5. HEC-RAS Output for 8.5 ft CMP at QP.

Plan: 8.5 1.55 0.045   North Thompson C   Tributary RS: 260   Culv Group: Culvert #1   Profile: Q25

Q Culv Group (cfs)	103.00	Culv Full Len (ft)	
# Barrels	1	Culv Vel US (ft/s)	6.25
Q Barrel (cfs)	103.00	Culv Vel DS (ft/s)	4.88
E.G. US. (ft)	96.58	Culv Inv El Up (ft)	91.63
W.S. US. (ft)	96.04	Culv Inv El Dn (ft)	90.40
E.G. DS (ft)	95.00	Culv Frctn Ls (ft)	0.92
W.S. DS (ft)	94.23	Culv Exit Loss (ft)	0.00
Delta EG (ft)	1.59	Culv Entr Loss (ft)	0.30
Delta WS (ft)	1.81	Q Weir (cfs)	
E.G. IC (ft)	95.80	Weir Sta Lft (ft)	
E.G. OC (ft)	96.22	Weir Sta Rgt (ft)	
Culvert Control	Outlet	Weir Submerg	
Culv WS Inlet (ft)	95.31	Weir Max Depth (ft)	
Culv WS Outlet (ft)	94.63	Weir Avg Depth (ft)	
Culv Nml Depth (ft)	3.66	Weir Flow Area (sq ft)	
Culv Crt Depth (ft)	3.39	Min El Weir Flow (ft)	101.67

Errors Warnings and Notes

Warning:	During subcritical analysis, the water surface upstream of culvert went to critical depth.
Note:	During the supercritical calculations a hydraulic jump occurred at the inlet of (going into) the
	culvert.

Table 1 - Culvert Summary Table: 8.5' CMP emb=2.55' n=0.060

Total Discharge (cfs)	Culvert Discharge (cfs)	Headwater Elevation (ft)	Inlet Control Depth (ft)	Outlet Control Depth (ft)	Flow Type	Normal Depth (ft)	Critical Depth (ft)	Outlet Depth (ft)	Tailwater Depth (ft)	Outlet Velocity (ft/s)	Tailwater Velocity (ft/s)
0.00	0.00	94.18	0.000	-1.230	0-NF	0.000	0.000	0.000	0.000	0.000	0.000
0.88	0.88	94.27	0.042	0.094	3-M1t	0.042	0.026	0.162	0.162	0.718	0.859
1.00	1.00	94.28	0.048	0.103	3-M1t	0.048	0.029	0.170	0.170	0.776	0.902
2.64	2.64	94.37	0.123	0.191	3-M1t	0.126	0.077	0.256	0.256	1.343	1.296
3.52	3.52	94.42	0.163	0.238	3-M1t	0.167	0.103	0.294	0.294	1.548	1.421
4.40	4.40	94.46	0.201	0.282	3-M1t	0.209	0.129	0.329	0.329	1.722	1.515
5.28	5.28	94.50	0.239	0.323	3-M1t	0.251	0.155	0.361	0.361	1.880	1.594
6.16	6.16	94.56	0.277	0.375	3-M1t	0.293	0.181	0.389	0.389	2.027	1.664
7.04	7.04	94.60	0.314	0.419	3-M1t	0.335	0.207	0.416	0.416	2.165	1.727
7.92	7.92	94.64	0.350	0.465	3-M1t	0.377	0.232	0.440	0.440	2.295	1.784
8.80	8.80	94.70	0.386	0.516	3-M1t	0.418	0.258	0.464	0.464	2.419	1.836

```
****************************************************************************
```

Inlet Elevation (invert): 94.18 ft,   Outlet Elevation (invert): 92.95 ft

Culvert Length: 46.02 ft,   Culvert Slope: 0.0267

```
****************************************************************************
```

Site Data - 8.5' CMP emb=2.55' n=0.060

Site Data Option: Culvert Invert Data

Inlet Station: 0.00 ft

Inlet Elevation: 91.63 ft

Outlet Station: 46.00 ft

Outlet Elevation: 90.40 ft

Number of Barrels: 1

Culvert Data Summary - 8.5' CMP emb=2.55' n=0.060

Barrel Shape: Circular

Barrel Diameter: 8.50 ft

Barrel Material: Corrugated Steel

Embedment: 30.60 in

Barrel Manning's n: 0.0240 (top and sides)

Manning's n: 0.0600 (bottom)

Inlet Type: Conventional

Inlet Edge Condition: Square Edge with Headwall

Inlet Depression: None

Table 2 - Downstream Channel Rating Curve (Crossing: N Thompson, n=0.060)

Flow (cfs)	Water Surface Elev (ft)	Depth (ft)	Velocity (ft/s)	Shear (psf)	Froude Number
0.00	92.95	0.00	0.00	0.00	0.00
0.88	93.11	0.16	0.86	0.27	0.48
1.00	93.12	0.17	0.90	0.28	0.49
2.64	93.21	0.26	1.30	0.42	0.53
3.52	93.24	0.29	1.42	0.49	0.55
4.40	93.28	0.33	1.51	0.55	0.56
5.28	93.31	0.36	1.59	0.60	0.56
6.16	93.34	0.39	1.66	0.65	0.57
7.04	93.37	0.42	1.73	0.69	0.57
7.92	93.39	0.44	1.78	0.73	0.58
8.80	93.41	0.46	1.84	0.77	0.58

Tailwater Channel Data - N Thompson, n=0.060

Tailwater Channel Option: Irregular Channel

Roadway Data for Crossing: N Thompson, n=0.060

Roadway Profile Shape: Constant Roadway Elevation

Crest Length: 99.00 ft

Crest Elevation: 101.66 ft

Roadway Surface: Paved

Roadway Top Width: 23.00 ft

H.3.7. HEC-RAS Output for 8.5 ft CMP with Oversized Bed at QH.

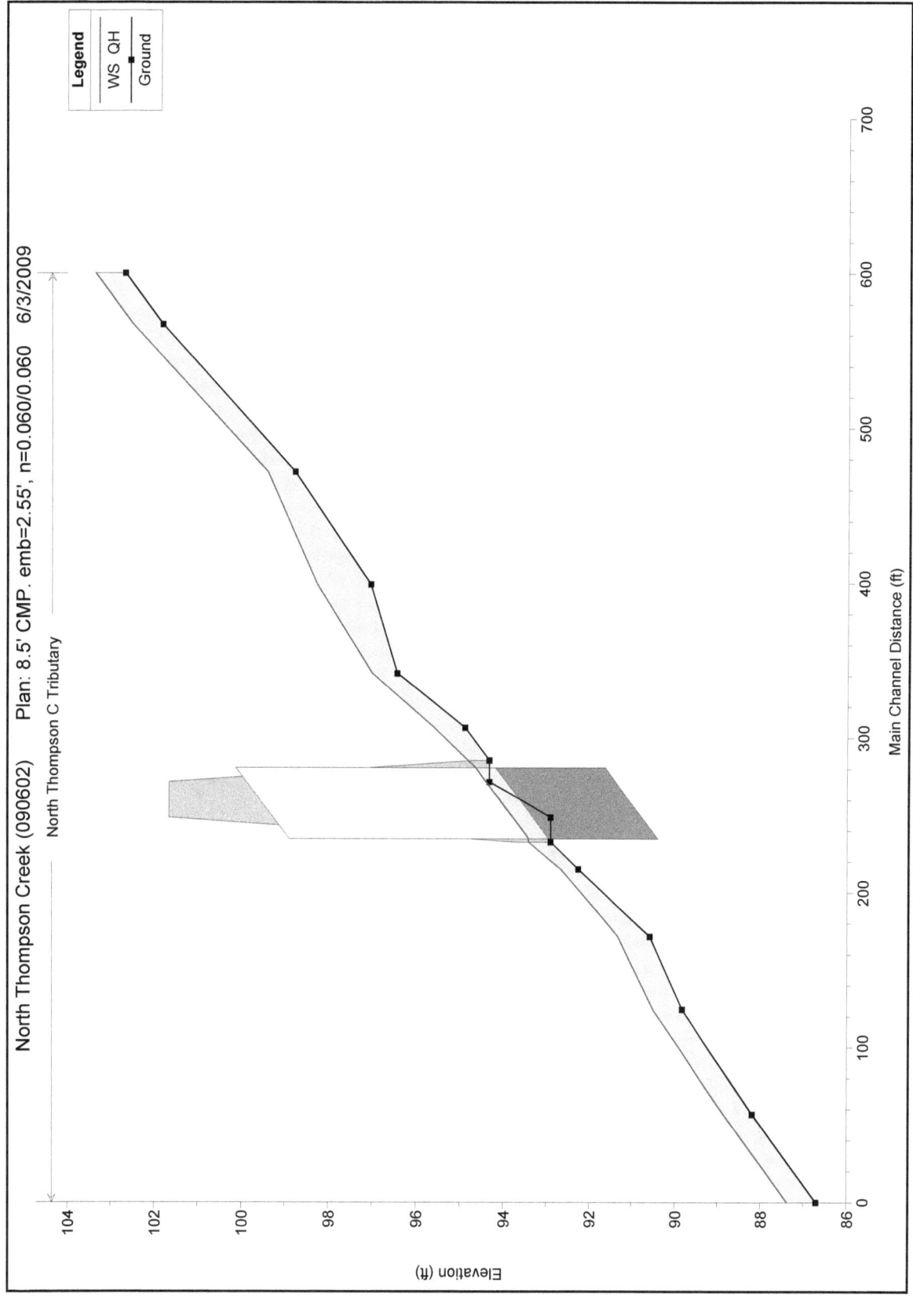

North Thompson Creek (090602)   Plan: 8.5' CMP. emb=2.55', n=0.060/0.060   6/3/2009
North Thompson C Tributary

H.3.7. HEC-RAS Output for 8.5 ft CMP with Oversized Bed at QH.

HEC-RAS Plan: 8.5 2.55 0.060   River: North Thompson C   Reach: Tributary   Profile: QH

Reach	River Sta	Profile	Q Total (cfs)	Min Ch El (ft)	W.S. Elev (ft)	Max Chl Dpth (ft)	Hydr Depth (ft)	Mann Wtd Chnl	Crit W.S. (ft)	E.G. Elev (ft)	E.G. Slope (ft/ft)	Vel Chnl (ft/s)	Flow Area (sq ft)	Top Width (ft)	Froude # Chl
Tributary	600	QH	8.80	102.69	103.37	0.68	0.44	0.057	103.25	103.46	0.026619	2.43	3.62	8.31	0.65
Tributary	567	QH	8.80	101.83	102.52	0.69	0.44	0.056		102.61	0.025371	2.41	3.66	8.32	0.64
Tributary	472	QH	8.80	98.79	99.42	0.63	0.43	0.057	99.35	99.55	0.041605	2.93	3.00	7.02	0.79
Tributary	399	QH	8.80	97.05	98.28	1.23	0.49	0.049	97.97	98.32	0.008884	1.69	5.20	10.63	0.43
Tributary	342	QH	8.80	96.44	97.01	0.57	0.36	0.057	97.01	97.20	0.071477	3.46	2.54	6.97	1.01
Tributary	307	QH	8.80	94.88	95.60	0.72	0.50	0.060	95.38	95.66	0.017209	2.01	4.37	8.75	0.50
Tributary	286	QH	8.80	94.32	94.82	0.50	0.33	0.060	94.82	94.98	0.077646	3.29	2.67	8.00	1.00
Tributary	260		Culvert												
Tributary	233	QH	8.80	92.90	93.42	0.52	0.42	0.056	93.30	93.50	0.021951	2.19	4.01	16.15	0.60
Tributary	215	QH	8.80	92.26	92.67	0.41	0.29	0.060	92.67	92.81	0.080273	3.04	2.90	10.09	1.00
Tributary	172	QH	8.80	90.59	91.32	0.73	0.55	0.056	91.09	91.38	0.013199	1.98	4.45	8.10	0.47
Tributary	125	QH	8.80	89.83	90.51	0.68	0.41	0.056		90.58	0.022114	2.16	4.08	9.94	0.59
Tributary	57	QH	8.80	88.20	88.86	0.66	0.48	0.057	88.73	88.96	0.026202	2.52	3.49	7.23	0.64
Tributary	0	QH	8.80	86.71	87.38	0.67	0.48	0.057	87.25	87.48	0.026048	2.52	3.49	7.23	0.64

H.3.7. HEC-RAS Output for 8.5 ft CMP with Oversized Bed at QH.

Plan: 8.5 2.55 0.060   North Thompson C   Tributary RS: 260   Culv Group: Culvert #1   Profile: QH

Q Culv Group (cfs)	8.80	Culv Full Len (ft)	
# Barrels	1	Culv Vel US (ft/s)	2.40
Q Barrel (cfs)	8.80	Culv Vel DS (ft/s)	2.34
E.G. US. (ft)	94.77	Culv Inv El Up (ft)	91.63
W.S. US. (ft)	94.82	Culv Inv El Dn (ft)	90.40
E.G. DS (ft)	93.50	Culv Frctn Ls (ft)	1.22
W.S. DS (ft)	93.42	Culv Exit Loss (ft)	0.01
Delta EG (ft)	1.28	Culv Entr Loss (ft)	0.04
Delta WS (ft)	1.39	Q Weir (cfs)	
E.G. IC (ft)	94.61	Weir Sta Lft (ft)	
E.G. OC (ft)	94.77	Weir Sta Rgt (ft)	
Culvert Control	Outlet	Weir Submerg	
Culv WS Inlet (ft)	94.64	Weir Max Depth (ft)	
Culv WS Outlet (ft)	93.42	Weir Avg Depth (ft)	
Culv Nml Depth (ft)	3.01	Weir Flow Area (sq ft)	
Culv Crt Depth (ft)	2.89	Min El Weir Flow (ft)	101.67

Errors Warnings and Notes

Warning:	During subcritical analysis, the water surface upstream of culvert went to critical depth.

H.4 REFERENCES

Blakemore, E. Thomas, H.W. Hjalmarson, an$_d$ S.D. Waltemeyer, 1997. "Methods for Estimating Magnitude and Frequency of Floods in the Southwestern United States," USGS Water Supply Paper 2433.

Kircher, James E., Anne F. Choquette, and Brian D. Richter, 1985. "Estimation of Natural Streamflow Characteristics in Western Colorado, USGS Water-Resources Investigations Report 85-4086.

Vaill, J.E., 2000. "Analysis of the Magnitude and Frequency of Floods in Colorado," USGS Water-Resources Investigations Report 99-4190.

APPENDIX I- DESIGN EXAMPLE: TRIBUTARY TO BEAR CREEK, ALASKA

I.1 SITE DESCRIPTION

The design procedure is applied to a road crossing of a Tributary to Bear Creek, which is approximately 12 miles (19.3 kilometers) south and 5 miles (8.1 kilometers) east of Petersburg, Alaska. The drainage area to the crossing is 0.23 mi^2 (0.60 km^2). The watershed is densely forested with many wetlands. Activities in the watershed include timber harvest. Elevations in the watershed range from 420 to 2300 ft (130 to 700 m). (All data and photos for this application were provided by Mark Weinhold of the USFS.)

There is an existing 60-in (1520-mm) culvert at the stream-road crossing. Figure I.1 shows the outlet of the culvert. The culvert is targeted for replacement because the USFS determined it to be a passage barrier, possibly because of the drop at the outlet and the velocity in the barrel.

Figure I.1.   Tributary to Bear Creek Culvert Outlet.

I.2 DESIGN PROCEDURE APPLICATION

This section illustrates the application of the design procedure. The uses of two separate tool sets are shown: 1) HY-8 with normal depth computations for the channel cross-sections and 2) HEC-RAS. Although both tool sets are shown, the designer may choose one or the other as appropriate for the site and the designers modeling skills.

Step 1. Determine Design Flows.

Discharges are determined for the peak flow, Q_P, high passage flow, Q_H, and low passage flow, Q_L. For the peak flow, Q_p, the Alaska Department of Transportation and Public Facilities (ADOT&PF) standard for culvert design are applied. Applicable ADOT&PF standards for culverts on roads such as the subject road range from a 50-yr event for "culverts on secondary highways providing sole area access" to a 10-yr event for "culverts on secondary highways of less importance." The more conservative 50-yr standard will be applied in this example.

Curran, et al. (2003) provides a set of USGS regression equations applicable to southeast Alaska. However, the equations for southeast Alaska have a minimum drainage area of 0.72 mi^2 (1.86 km^2), which is well above the project drainage area of 0.23 mi^2 (0.60 km^2). Another hydrologic method recommended by ADOT&PF (1995) is the SCS unit hydrograph. This method as implemented in WinTR-55, was applied to compute the 50-yr discharge. Selected characteristics required by one or both of these methods are summarized in Table I.1. The results for the SCS and regression approaches are summarized in Table I.2. The SCS approach results in a higher Q_{50}, but a lower Q_2 compared with the regression equations. Since the regression equation is not applicable, a discharge of 216 ft^3/s (6.1 m^3/s) will be used for Q_p.

Table I.1.   Watershed and Rainfall Characteristics.

Characteristic	CU	SI
Drainage area	0.23 mi^2	0.60 km^2
Percent lakes and ponds	0 %	0 %
Mean annual precipitation	100 in	2540 mm
Mean minimum January temperature	26 degrees F	-3 degrees C
Curve number	55	55
Time of concentration	1 h	1 h

Table I.2.   Discharge Estimates.

Discharge Quantity	Curran, et al. (2003) Region 1, ft^3/s (m^3/s)	SCS Unit Hydrograph (WinTR-55), ft^3/s (m^3/s)	Wiley and Curran (2003) Region 1, ft^3/s (m^3/s)
Q_{50}	144 (4.08)	Q_P = 216 (6.1)	
Q_2	61 (1.7)	25 (0.71)	
$Q_{10\%}$			4.7 (0.13)
$0.25Q_2$	15 (0.42)	6.2 (0.18)	
$0.4Q_2$	Q_H = 24 (0.68)	10 (0.28)	
$Q_{90\%}$			0.23 (0.0065)
7Q2			0.21 (0.0059)
Q_L (min)			Q_L = 1.0 (0.028)

The high passage flow is determined by site-specific guidelines, if they exist. Gubernick (1995) has suggested that for southeast Alaska the high passage flow may be estimated as 40 percent of the Q_2. These values are given in Table I.2. In the absence of site-specific guidelines, the Q_H may be defined as the 10 percent exceedance quantile on the annual flow duration curve. A flow duration curve does not exist for this location, but Wiley and Curran (2003) include a regression equation that estimates a 10 percent exceedance flow. However, the minimum drainage area for these equations exceeds the drainage area of our site.

Table I.2 summarizes the high passage flow estimates available ranging from 6.2 ft^3/s (0.18 m^3/s) (25 percent of the SCS Q_2) and 4.7 ft^3/s (0.13 m^3/s) (Wiley and Curran 10 percent exceedance) to 24 ft^3/s (0.68 m^3/s) (40 percent of the Curran, et al. Q_2). Because the watershed is less than the minimum drainage area for the Curran et al. and Wiley and Curran based estimates, the SCS based estimate using the guidance from Gubernick suggests that the high passage flow should be 10 ft^3/s (0.28 m^3/s). However, given the wide range of estimates, the high passage flow, Q_H, will be taken to be 24 ft^3/s (0.68 m^3/s) to be conservative in representing the flows for which passage should be considered.

The low passage flow is determined by site-specific guidelines, if they exist. None are known to exist for this site. In the absence of site-specific guidelines, the Q_L should be defined as the 90 percent exceedance quantile on the annual flow duration curve or the 7-day, 2-yr low flow (7Q2), but no less than 1 ft³/s (0.028 m³/s). As previously noted, a flow duration curve does not exist for this location, but Wiley and Curran (2003) include regression equations for both the 90 percent exceedance flow and the 7Q2 for the July – September period. These values are summarized in Table I.2. However, both of these estimates are less than 1 ft³/s (0.028 m³/s), which will be taken as Q_L.

It should be noted that the natural channel is step-pool in form and may not be passable at low flows. Additional consideration of an appropriate low flow could result in a higher value for Q_L.

Step 2. Determine Project Reach and Representative Channel Characteristics.

The project reach should extend no less than three culvert lengths or 200 ft (61 m), whichever is greater, up and downstream of the crossing location. Since the existing culvert is 51 ft (15.5 m) in length, the project reach must extend at least 200 ft upstream and downstream of the culvert inlet and outlet, respectively. At least three cross-sections should be obtained upstream and downstream from the crossing location.

Five stream cross-sections were collected by the USFS. Two are downstream of the culvert; the most downstream cross-section is approximately 80 ft (24 m) downstream from the road centerline. To provide for a third downstream cross-section and extend the project reach in the downstream direction, an additional cross section was created by copying the surveyed cross-section at station 88 and lowering it such that the cross-section thalweg matched the surveyed stream profile at that location.

Three of the USFS cross-sections are upstream; the most upstream cross-section is approximately 220 ft (67 m) upstream of the road centerline. Table I.3 summarizes the cross-section locations and Figure I.2 shows the creek and cross-sections schematically in plan view. Cross-sections shown in Figure I.2, but not listed in Table I.3 were interpolated for the purpose of water surface profile modeling with HEC-RAS. Plots of the surveyed cross-sections are included in Section I.3.1.

Table I.3. Surveyed Cross-Sections.

Cross-section	Station (ft)	Station (m)	Thalweg Elevation (ft)	Thalweg Elevation (m)	Slope to downstream cross-section (ft/ft or m/m))
394	394	120	334.31	101.90	0.096
374	374	114	332.39	101.31	0.080
333	333	101	329.13	100.32	0.061
Road centerline	172	52	--	--	--
101	101	31	310.10	94.52	0.128
88	88	27	308.44	94.01	0.088
0 (estimated)	0	0	300.69	91.65	--

The longitudinal profile of the stream and existing roadway embankment is shown in Figure I.3 using all surveyed data collected by the USFS. Although only the data acquired at the cross-section locations are used in the analysis, the longitudinal detail shows the variability within the natural stream. Superimposed on the detailed profile are the cross-section locations plotted with their thalweg elevations as well as the existing culvert invert. The existing culvert slope is 3.3 percent.

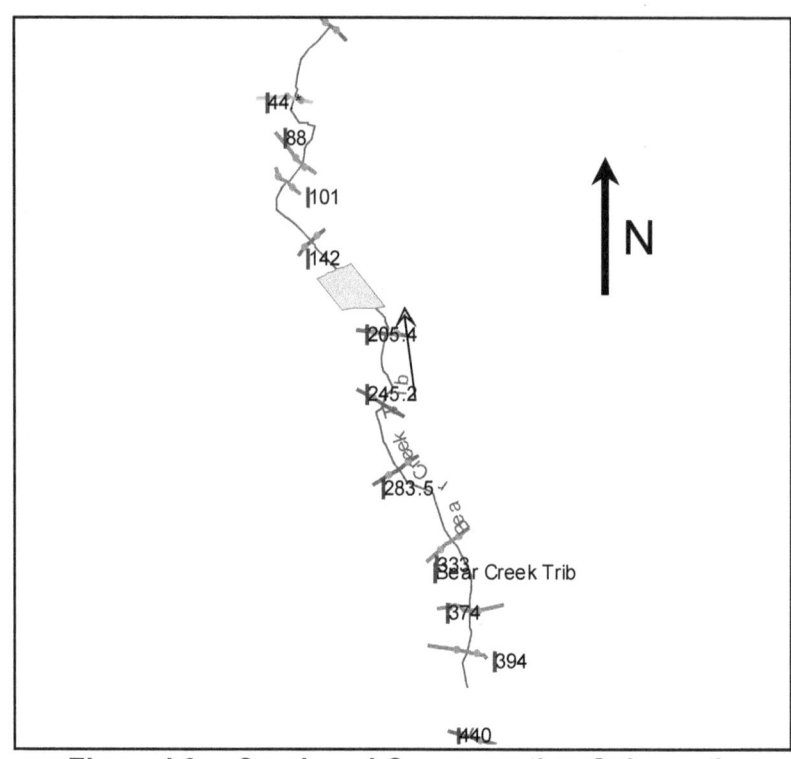

Figure I.2.   Creek and Cross-section Schematic.

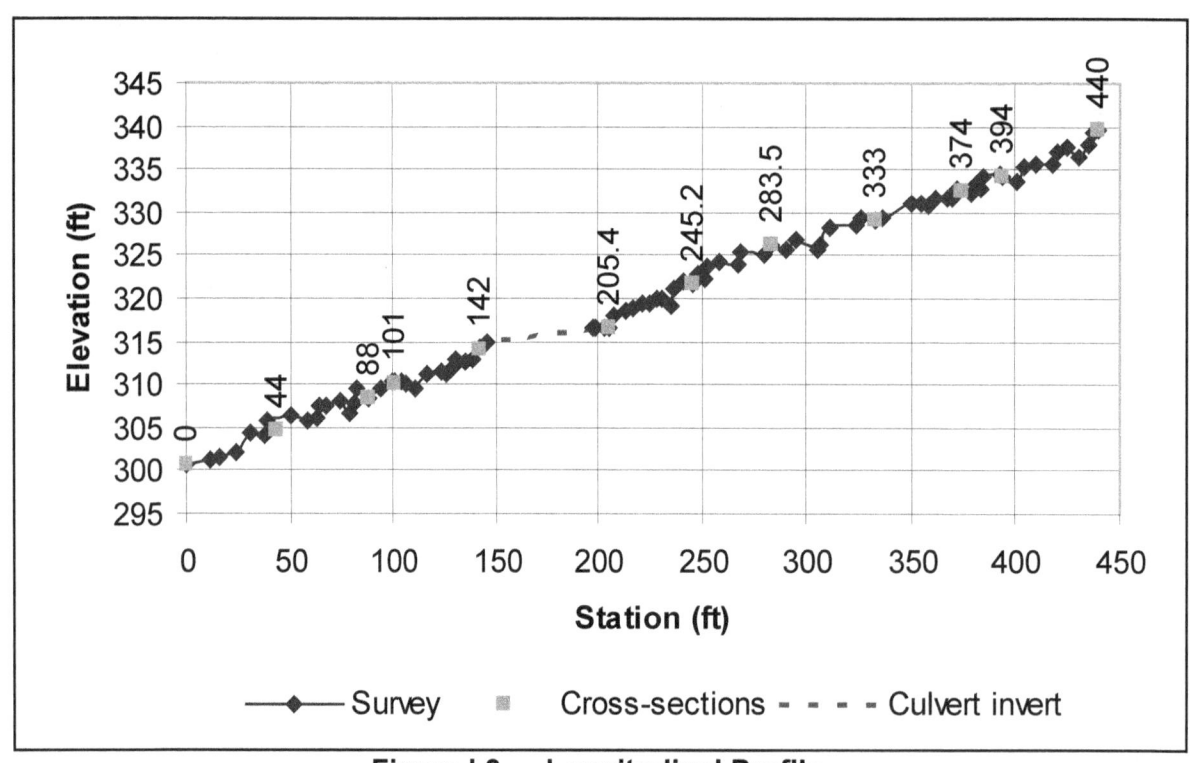

Figure I.3.   Longitudinal Profile.

One bed material gradation was collected in the project reach. A random walk on 3 ft (0.91 m) intervals was conducted between cross-sections 333 and 374 to generate the pebble count. Although two samples are preferred as a minimum, one upstream and one downstream, only one was collected for this site because it was considered the appropriate material for the culvert bed. The gradation is summarized in Figure I.4 and Table I.4. Evidence of bed armoring was not reported. The unit weight of the bed material was not provided so a value of 156 lb/ft^3 (24,500 N/m^3) was assumed.

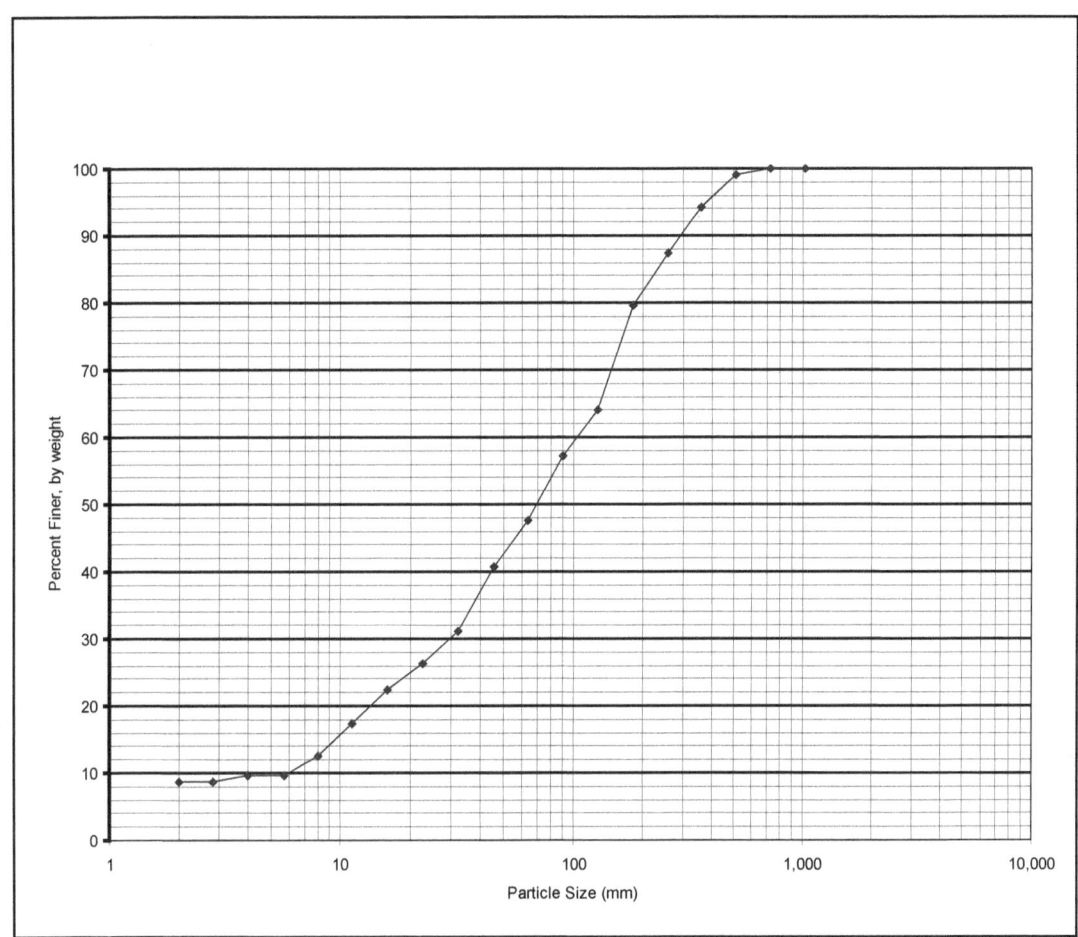

Figure I.4.   Bed Material Gradation.

Table I.4.   Bed Material Quantiles.

Quantile	Size (ft)	Size (mm)
D_{95}	1.27	387
D_{84}	0.73	223
D_{50}	0.23	71
D_{16}	0.032	10
D_5	0.007	2

Step 3. Check for Dynamic Equilibrium.

The qualitative assessment for dynamic equilibrium involves three components:

1. Watershed reconnaissance for changes in supply.

2. Project reach sediment transport assessment.

3. Field observations of the project reach.

The watershed reconnaissance is to identify changes in the watershed that may result in changes in sediment supply. As described in the site summary, activities in the watershed include timber harvest. Timber removed from the watershed will be replaced with new trees over time and other uses are not reported to be intensive and are not forecast to change significantly.

The project reach sediment transport assessment examines potential changes in discharge, slope, and D_{50} throughout the project reach in the context of Lane's proportional relationship to sediment transport (Equation 7.2). Discharge throughout the reach is invariant and the sediment size distribution is assumed to be relatively constant. Average slopes in the project reach range from 0.06 ft/ft to 0.13 ft/ft (see Table I.3). Slope variations are not considered to be an issue for sediment transport.

Field observations are to identify any indicators of instability. No instabilities were reported.

Taking the components together, there is no clear concern that the project reach is experiencing instability or disequilibrium in sediment transport.

Step 4. Analyze and Mitigate Channel Instability.

Based on the assessment in Step 3, this step is unnecessary.

Step 5. Align and Size Culvert for Q_p.

ADOT&PF criteria allow for a headwater depth to culvert rise ratio (HW/D ratio) of up to 1.5 provided the headwater does not damage upstream property and the flow is not diverted away from the culvert. An embedded CMP culvert will be designed with these criteria. If woody debris is a concern, the allowable HW/D ratio should be lower. Woody debris is not reported as a particular concern for this site.

Since there is an existing CMP culvert (5-ft diameter on a slope of 3.3 percent) this size and vertical alignment is analyzed as a starting point. For the peak design flow, Q_P, of 216 cfs the existing culvert (without embedment) has a HW/D ratio of 1.7 and overtops the roadway. Therefore, the existing culvert does not meet hydraulic criteria.

The horizontal alignment of the existing culvert will be maintained.

The desired vertical alignment of the replacement culvert is established by evaluation of the vertical profile of the stream and the existing culvert profile. The profile in Figure I.3 averages 0.085 ft/ft in slope. For the initial design trial, the culvert will be laid out more steeply than the existing culvert, but modestly less than the average slope. The initial slope will be 0.079 ft/ft, which was selected because a culvert on this slope will tie in with control points on the stream channel. A rationale for increasing the culvert slope is to reduce the overly steep channel profile immediately up and downstream of the existing culvert and to provide a culvert slope that more closely parallels the natural stream channel. This will require lowering the outlet 1.4 ft (and eliminating the jump required to enter the culvert (See Figure I.1)) and raising the inlet 1.0 ft. During installation of the culvert, these adjustments will be made in conjunction with adjusting the adjacent stream profile.

An initial CMP culvert diameter of 6.5 ft is estimated considering that the existing 5-ft diameter culvert is inadequate, and that a 2 ft minimum embedment is required. The embedment criteria for a circular culvert are 30 percent of the culvert rise giving an embedded depth of 0.3 x 6.5 ft = 1.95 ft. However, embedment depth may be no less than 2.0 ft or 2 times the D_{95} of the bed material. From Table I.4, the D_{95} is 1.27 ft thereby requiring the minimum embedment depth to be 2.6 ft for a 6.5-ft CMP. Inlet and outlet elevations for the existing and replacement culvert are summarized in Table I.5.

Table I.5.   Inlet and Outlet Elevations for Existing and Replacement Culverts.

Description	Inlet	Outlet
Existing Culvert Invert	316.6	314.9
Replacement Culvert Bed	317.6	313.5
Replacement Culvert Invert	315.0	310.9

A bed gradation must be selected. Since we have a single gradation sample from the site, that gradation will be selected for design as is shown in Table I.6. Recall that to control bed interstitial flow, it is recommended that the D_5 fraction be no larger than 2 mm (sand, silt, and clay). The existing gradation satisfies this requirement; no modification is required.

Table I.6.   Bed Gradation Design.

Quantile	Design (mm)	Design (ft)
D_{95}	387	1.27
D_{84}	223	0.73
D_{50}	71	0.23
D_{16}	10	0.033
D_5	2	0.0066

A Manning's n is needed to estimate the roughness of the bed material in the culvert. The Limerinos equation is used over a range of depths using the depth above the bed rather than of the hydraulic radius (see Appendix C). Since we do not know the flow depth, the roughness value will be calculated over a range of depths. The calculation for a depth of 1 ft is as follows. Table I.7 summarizes Manning's n for a range of depths.

$$n = \frac{\alpha \, y^{1/6}}{1.16 + 2\log\left(\dfrac{y}{D_{84}}\right)} = \frac{0.0926(1.0)^{1/6}}{1.16 + 2\log\left(\dfrac{1.0}{0.73}\right)} = 0.065$$

Table I.7.   Manning's n for Bed Material using the Limerinos Equation.

Depth (ft)	Manning's n
1.0	0.065
1.5	0.056
2.0	0.051
2.5	0.048
3.0	0.047
3.5	0.045
4.0	0.044

The Manning's n corresponding with normal depth in the culvert at the design flow is used for the bed material in this step. Since normal depth is a function of Manning's n, an iterative process is required to determine the normal depth and Manning's n. For a 6.5 ft CMP culvert with 2.6 ft of embedment the normal depth at Q_P is 2.74 ft with a Manning's n of 0.048 (Table I.7). For these conditions the culvert operates in inlet control with a headwater depth of 8.33 ft. The HW/D ratio is 8.33/(6.5-2.6) = 2.1, which is exceeds the 1.5 maximum criteria.

Next, try a 7.5 ft CMP. In this case, the culvert operates in inlet control with a headwater depth of 6.42 ft. The HW/D ratio is 6.42/(7.5-2.6) = 1.3, which is less than the 1.5 maximum criteria. This headwater also does not overtop the road or result in a redirection of flows away from the culvert. Although this culvert has a higher rise than the existing culvert, minimum cover still appears to be available. This culvert alignment, size, and type are adopted as we progress to the next step.

Step 6. Check Culvert Bed Stability at Q_H.

Since the proposed culvert bed is greater than 5 percent, the critical unit discharge approach will be applied. The unit discharge in the culvert is compared with the critical unit discharge for the bed material.

The critical unit discharge is computed using Equation 7.12. However, we first use Equation 7.11 to determine the critical unit discharge for uniform materials:

$$q_{c-D50} = \frac{0.15g^{0.5}D_{50}^{1.5}}{S^{1.12}} = \frac{0.15(32.2)^{0.5}(0.23)^{1.5}}{(0.079)^{1.12}} = 1.61 \text{ ft}^3/\text{s/ft}$$

Using Equation 7.13 to compute the exponent b we can then use Equation 7.12 to calculate the critical unit discharge for the D_{84}:

$b = 1.5(D_{84}/D_{16})^{-1} = 1.5(0.73/0.033)^{-1} = 0.068$

$q_{c-D84} = q_{c-D50} (D_{84}/D_{50})^b = 1.61(0.73/0.23)^{0.068} = 1.74 \text{ ft}^2/\text{s}$

To compute the unit discharge in the culvert we need to determine the active channel bed width, which will depend on the culvert geometry, bed roughness, and discharge. Bed roughness is estimated iteratively assuming normal depth in the culvert.

The iteration is accomplished by assuming a normal depth, estimating the Manning's n based on that depth, and then applying HY-8 with that roughness to determine the normal depth. If the calculated depth matches the assumed depth, the iteration is completed. If not, a new depth is assumed and the process repeated. At Q_H, the Manning's n is 0.074 and normal depth in the culvert equals 0.77 ft.

HY-8 reports that the culvert is operating in outlet control with flow type 3-M1t at Q_H (see Section I.3.2). The inlet and outlet flow depths are taken from the water surface profile information in HY-8 as 0.79 ft and 1.05 ft, respectively. (The culvert summary table reports a slightly different depth at the outlet of 1.04 ft.) The active bed width, or water surface width in the case of the culvert, is calculated from the embedded culvert geometry. For the inlet, the calculations are:

y = 0.79 ft

w_a = 7.465 ft

q = 24/7.465 = 3.21 ft²/s

The results of the inlet and outlet calculation are summarized in Table I.8. Both inlet and outlet unit discharges exceed the critical unit discharge, therefore, the culvert bed is not considered stable at Q_H.

Performing the same analysis with HEC-RAS, yields inlet and outlet depths of 0.73 and 1.30 ft, respectively, with the culvert operating under outlet control. (See Section I.3.3.) The depths, velocities, and maximum shear stresses are summarized in Table I.8. Unit discharges based on these depths are also summarized in Table I.8. Both inlet and outlet unit discharges exceed the critical unit discharge, therefore, the culvert bed is not considered stable at Q_H.

Both methods lead to the conclusion that the bed is not stable at Q_H, therefore we must proceed to Step 7.

Table I.8.   7.5 ft CMP Culvert Inlet and Outlet Parameters at Q_H.

Parameter	HY-8		HEC-RAS	
	Inlet	Outlet	Inlet	Outlet
y (ft)	0.79	1.05	0.73	1.30
w_a (ft)	7.465	7.497	7.453	7.500
q (ft^2/s)	3.21	3.20	3.22	3.20
$q_{c\text{-}D84}$ (ft^2/s)	1.74	1.74	1.74	1.74

*Embedment=2.6 ft, S_o=0.079 ft/ft, n_{bed}=0.074.

Step 7. Check Channel Bed Mobility at Q_H.

The assessment in Step 6 concluded that the bed material in the culvert bottom is not stable at Q_H. In this step, we evaluate whether material is moving at this discharge in the upstream and the downstream channels. Table I.9 summarizes the unit discharge estimates at the cross-sections upstream and downstream using alternative methods: 1) a normal depth assumption and 2) HEC-RAS. The table also provides the unit discharges at the inlet and outlet of the culvert and the critical unit discharge for the bed material.

Table I.9.   Estimated Unit Discharges at Q_H.

Cross-section**	Normal Depth/HY-8 (ft^2/s)	HEC-RAS (ft^2/s)
394	2.59	2.53
374	3.03	3.06
333	2.95	2.92
Culvert Inlet*	3.21	3.22
Culvert Outlet*	3.20	3.20
101	3.57	3.51
88	3.57	3.59
$q_{c\text{-}D84}$ (ft^2/s)	1.74	1.74

*7.5 ft CMP, embedment = 2.6 ft, n_{bed} = 0.074.
**n_{bed} = 0.074.

In accordance with the guidance for this step, we observe whether or not the unit discharge in any channel cross-section is less than the critical unit discharge. The answer to this question is no, therefore, the bed is considered mobile. Next we review whether or not the unit discharge values in the culvert fall within the range of those estimated in the upstream and downstream channel cross-sections. Because the culvert values do fall within the range observed elsewhere in the project reach, the culvert bed material will move as the streambed material moves and we can proceed to Step 8.

Step 8. Check Culvert Bed Stability at Q_P.

The unit discharge in the culvert is compared with the critical unit discharge for the bed material. However, since the culvert unit discharges exceeded the critical unit discharge at Q_H, the unit discharges as Q_P will also exceed critical. Therefore, the check is not satisfied and we proceed to Step 9.

Step 9. Design Stable Bed for Q_P.

A stable bed design is attempted to resist the shear stresses at Q_P within the culvert. The bed will consist of a top layer of native material and an oversized underlayer. Design of the underlayer assumes the native top layer has been washed away at or before the peak of the hydrograph. It is assumed that natural replenishment cannot be relied on to restore the bed material in the culvert. If, however, site-specific analysis to the contrary is performed, the stable sublayer may be avoided.

As a first trial, select an oversized bed material that fits within the current culvert embedment of 2.6 ft. In accordance with the embedment criteria for Step 9, we would provide a 1 ft layer of native material leaving 1.6 ft for the oversized bed material. For a CMP culvert, the oversize layer minimum embedment is $1.5D_{95}$, therefore, D_{95} = 1.6 ft/1.5 = 1.1 ft. (Note that this quantile for the oversize bed is actually smaller than the D_{95} quantile in the native bed material.) Using the relation in Equation 7.15c between D_{50} and D_{95} for an oversized bed, the D_{50} = D_{95}/1.9 = 1.1 ft/1.9 = 0.58 ft. However, we will learn that this bed will not be stable at Q_P.

After an unsuccessful trial with a 9 ft CMP, consider a 12 ft CMP. Using the span as the active channel width results in a unit discharge of 216/12 = 18 ft^2/s. Setting this unit discharge as the critical unit discharge we can calculate a D_{50} that will be stable based on Equations 7.17 and 7.11. If we assume that D_{16} = $0.1D_{84}$, the stable D_{50} is 1.11 ft. (The relation between D_{84} and D_{16} will be reassessed prior to finalizing the design gradation.)

The thickness of the oversized layer is $1.5D_{95}$ and according to Equation 7.15c, D_{95} = $1.9D_{50}$. Therefore, the oversize layer thickness is 1.11 x 1.5 x 1.9 = 3.16 ft. In accordance with the embedment criteria, we would also provide a 1.27 ft layer of native material (D_{95} = 1.27 ft) above the oversize layer for a total embedment depth of 4.43 ft. The embedment represents 37 percent of the culvert rise, exceeding the minimum 30 percent, which is acceptable.

The D_5 is taken to be no larger than 2 mm to limit interstitial flow. The D_{16} was previously assumed to be $0.1D_{84}$. This represents a reasonable value for the oversize bed gradation. Table I.10 summarizes the resulting gradation and compares it to the native bed gradation.

Table I.10.   Oversize Stable Bed Design Gradation.

Quantile	Native (mm)	Oversize (mm)	Native (ft)	Oversize (ft)
D_{95}	387	640	1.27	2.1
D_{84}	223	457	0.73	1.5
D_{50}	71	335	0.23	1.1
D_{16}	10	46	0.033	0.15
D_5	2	2	0.0066	0.0066

With the native and oversize bed layers designed, proceed to Step 10.

Step 10. Check Culvert Velocity at Q_H.

A check is conducted to verify that the culvert velocities are less than or equal to those in at least part of the upstream or downstream channel. Table I.11 summarizes the velocities estimated at each cross-section and within the culvert by the HY-8/Normal Depth and HEC-RAS methods. (See sections I.3.4 and I.3.5, respectively.) The check is satisfied if the culvert inlet and outlet velocities are not the most severe in the project reach. For the both computations, the check is satisfied. However, the HEC-RAS cross-section computations are considered more accurate because they are based on a water surface profile calculation rather than assuming that the downstream bed profile represents the energy slope at the cross-section for the normal depth computation. Proceed to Step 11.

Table I.11.   Velocity Estimates at Q_H.

Cross-section**	Applicable Reach Length (ft)	Normal Depth/HY-8 (ft/s)	HEC-RAS (ft/s)
394	20	4.23	3.99
374	41	4.32	4.62
333	50	3.85	3.51
Culvert Inlet*	26	4.27	3.73
Culvert Outlet*	26	1.94	1.49
101	13	5.22	5.28
88	44	4.62	4.30

*12 ft CMP, embedment = 4.43 ft, n_b = 0.074.
**n_b = 0.074.

Step 11. Check Culvert Water Depth at Q_L.

A check is conducted to verify that the water depth in the culvert is greater than or equal to at least part of the upstream or downstream channel. Table I.12 summarizes the calculations for both methods. The water depths near the inlet in the culvert bed are shallower than in the upstream and downstream channel. Therefore, the check is not satisfied. Proceed to Step 12 to provide a low-flow channel.

Table I.12.   Maximum Depth Estimates at Q_L.

Cross-section**	Normal Depth/HY-8 (ft)	HEC-RAS (ft)
394	0.53	0.62
374	0.46	0.22
333	0.51	0.62
Culvert Inlet*	0.15	0.19
Culvert Outlet*	0.47	0.43
101	0.42	0.50
88	0.45	0.29

*12.0 ft CMP, embedment = 4.43 ft, n_b = 0.300.
**n_b = 0.300.

Step 12. Provide Low-flow Channel in Culvert.

To increase the depth in the culvert bed, add a triangular low-flow channel with side slopes of 1:8 (V:H). This will provide a thalweg 0.75 ft deeper in the center of the culvert. However, noting that the D_{84} of the native bed material is 0.73 ft the "construction" of such a small channel will require careful manual work.

Step 13. Review Design.

A 12.0-ft (3.66-m) CMP with a 4.43-ft (1.35-m) embedment on a 7.9 percent slope is proposed to replace the 5.0-ft (1.52-m) CMP culvert on a 3.3 percent slope. The average project reach slope is approximately 8.0 percent. An oversized underlayer is to be placed in the culvert to provide stability at Q_P. A low-flow channel is to be created to maintain depths in the culvert at Q_L.

Alternative culvert shapes and materials may also be considered. A concrete box or pipe arch may offer an option to maintain a sufficiently wide span to meet the stability, velocity, and depth criteria with a lower rise.

I.3   SUPPORTING DOCUMENTATION

I.3.1   Surveyed Cross-sections.

I.3.2   HY-8 Report for 7.5 ft CMP at Q_H.

I.3.3   HEC-RAS Output for 7.5 ft CMP at Q_H.

I.3.4   HY-8 Report for 12.0 ft CMP with oversized bed at Q_H.

I.3.5   HEC-RAS Output for 12.0 ft CMP with oversized bed at Q_H.

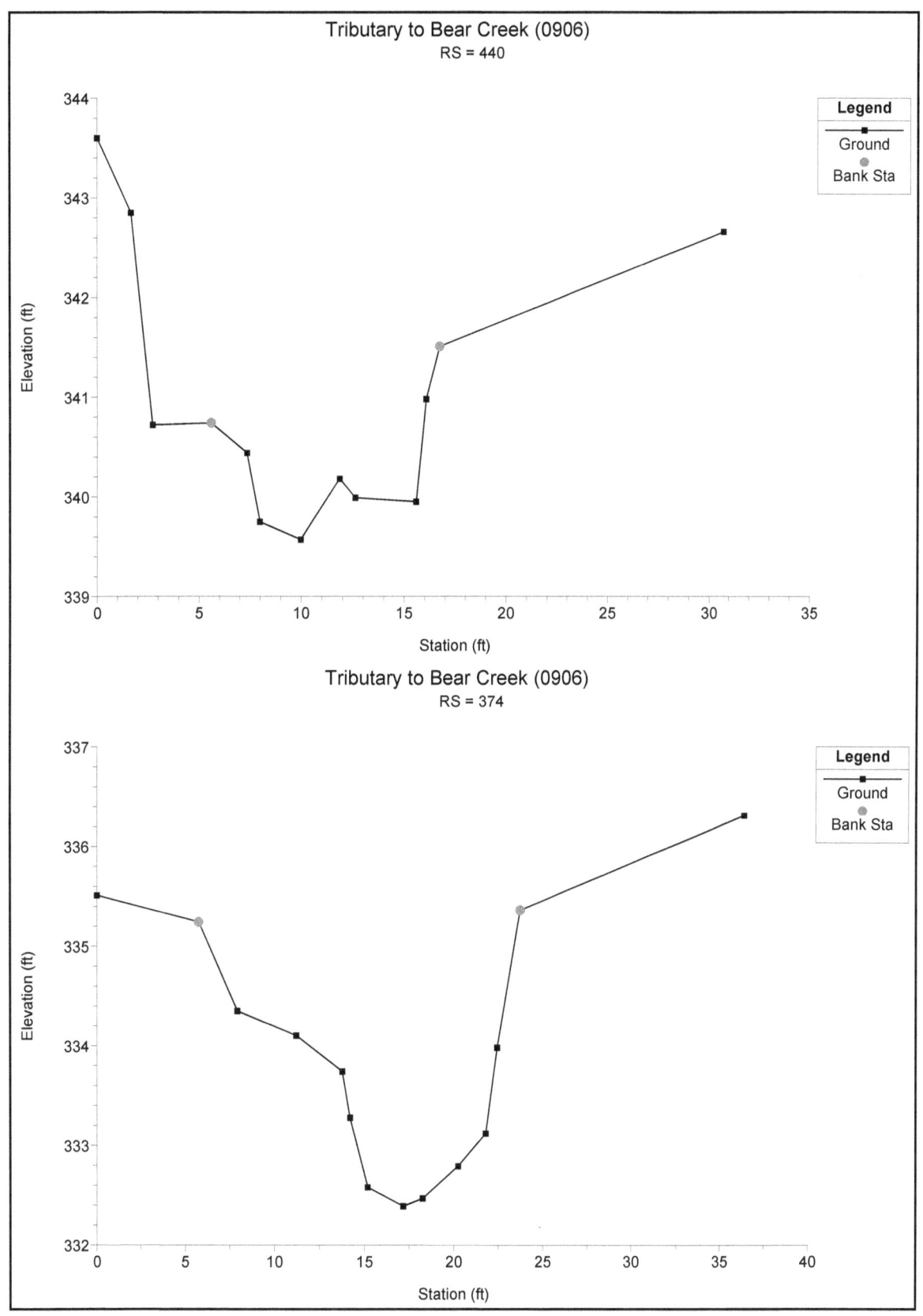

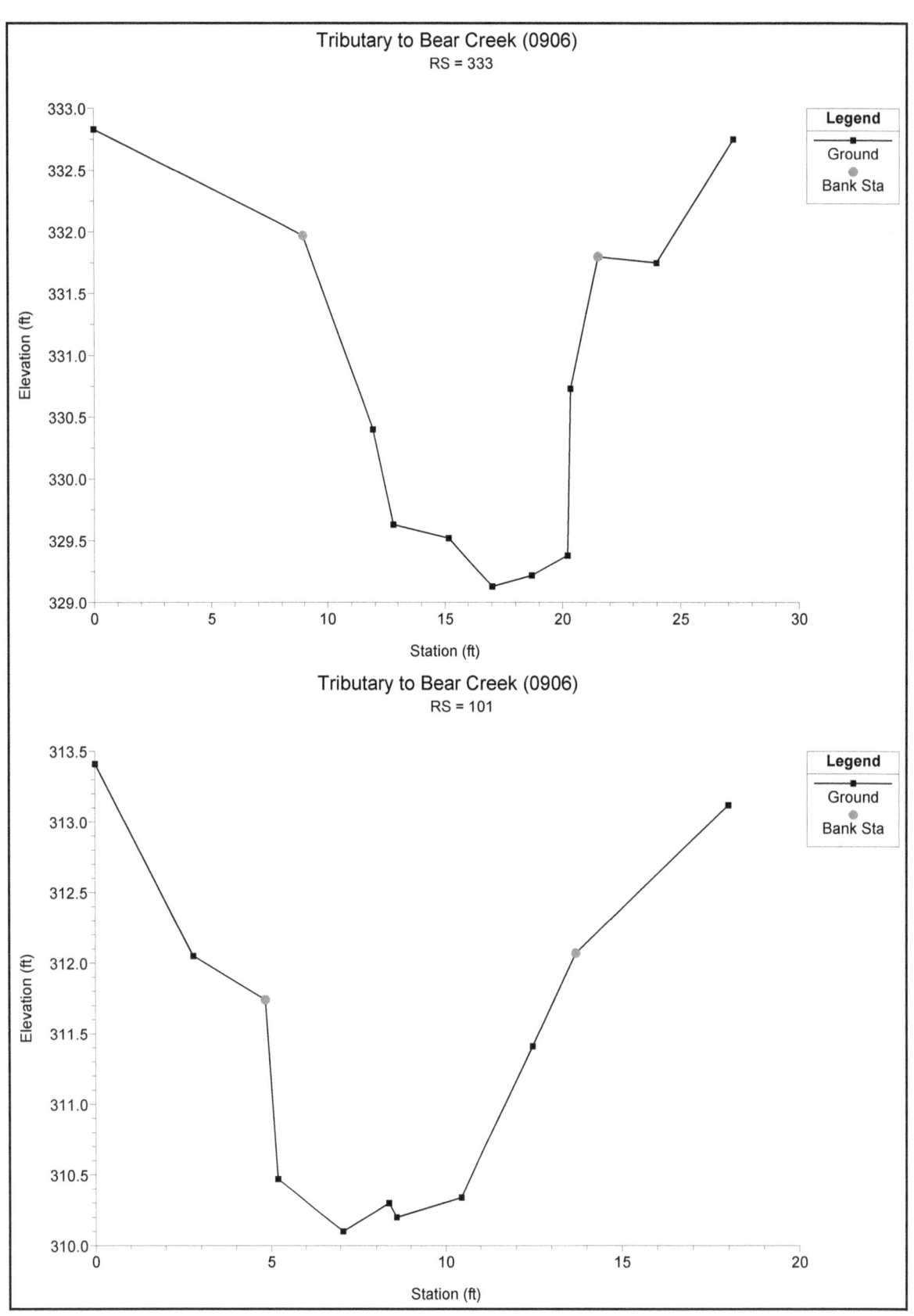

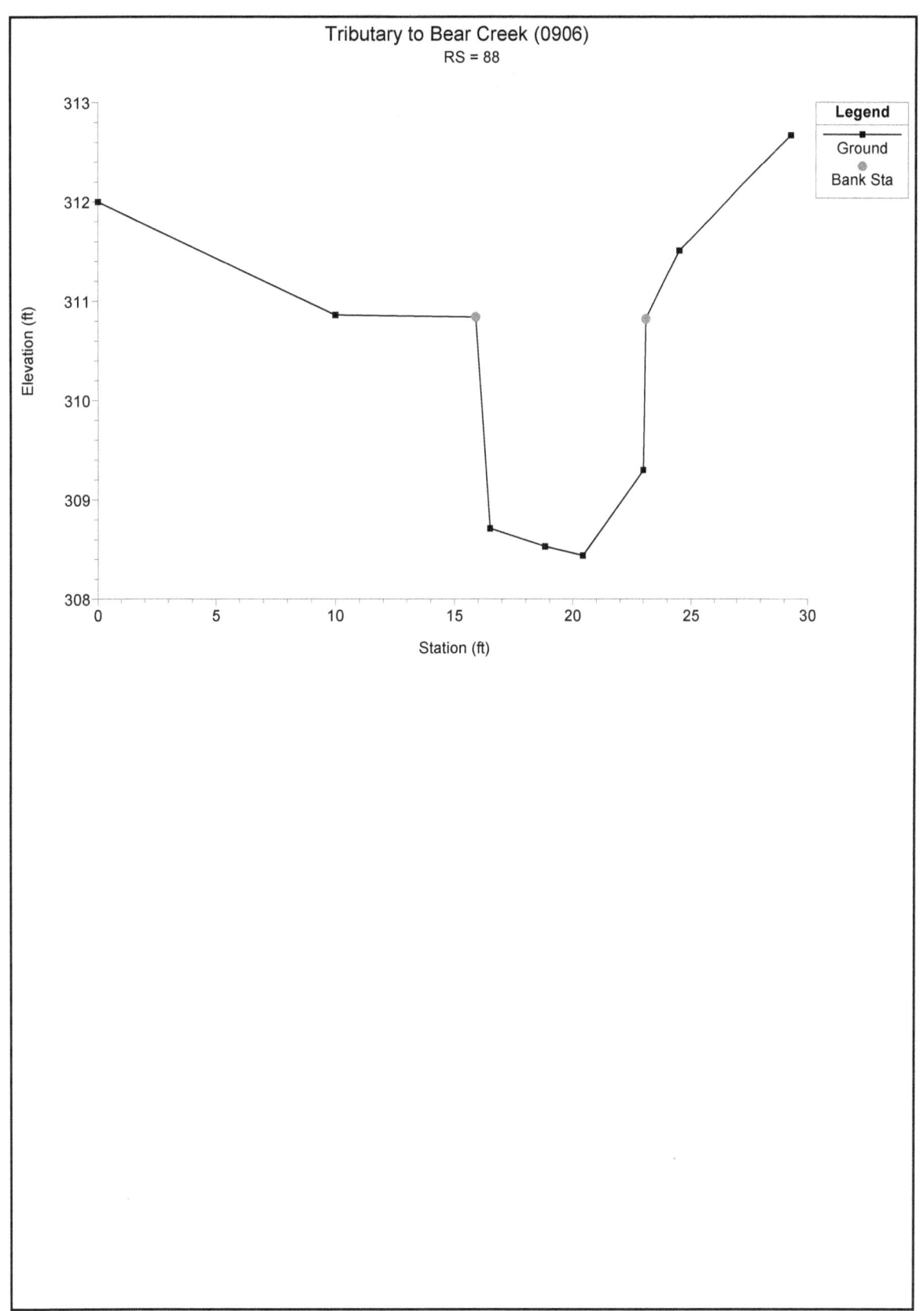

Tributary to Bear Creek (0906)
RS = 88

Table 1 - Culvert Summary Table: 7.5' CMP e=2.6' nb=0.074

Total Discharge (cfs)	Culvert Discharge (cfs)	Headwater Elevation (ft)	Inlet Control Depth (ft)	Outlet Control Depth (ft)	Flow Type	Normal Depth (ft)	Critical Depth (ft)	Outlet Depth (ft)	Tailwater Depth (ft)	Outlet Velocity (ft/s)	Tailwater Velocity (ft/s)
0.00	0.00	317.60	0.000	-4.100	0-NF	0.000	0.000	0.000	0.000	0.000	0.000
2.40	2.40	317.78	0.135	0.183	3-M1t	0.102	0.085	0.377	0.377	0.888	2.053
4.80	4.80	317.91	0.260	0.309	3-M1t	0.204	0.169	0.492	0.492	1.353	2.642
7.20	7.20	318.04	0.379	0.437	3-M1t	0.306	0.254	0.585	0.585	1.700	3.048
9.60	9.60	318.19	0.502	0.590	3-M1t	0.408	0.339	0.666	0.666	1.985	3.365
12.00	12.00	318.40	0.644	0.798	3-M1t	0.500	0.424	0.739	0.739	2.231	3.628
14.40	14.40	318.45	0.805	0.853	3-M1t	0.551	0.500	0.806	0.806	2.450	3.854
16.80	16.80	318.54	0.879	0.945	3-M1t	0.602	0.545	0.869	0.869	2.648	4.053
19.20	19.20	318.63	0.961	1.031	3-M1t	0.652	0.590	0.928	0.928	2.830	4.232
21.60	21.60	318.71	1.039	1.114	3-M1t	0.703	0.635	0.984	0.984	2.999	4.394
24.00	24.00	318.79	1.113	1.195	3-M1t	0.754	0.681	1.037	1.037	3.157	4.543

****

Inlet Elevation (invert): 317.60 ft,   Outlet Elevation (invert): 313.50 ft

Culvert Length: 52.06 ft,   Culvert Slope: 0.0790

****

Site Data - 7.5' CMP e=2.6' nb=0.074

Site Data Option: Culvert Invert Data

Inlet Station: 0.00 ft

Inlet Elevation: 315.00 ft

Outlet Station: 51.90 ft

Outlet Elevation: 310.90 ft

Number of Barrels: 1

Culvert Data Summary - 7.5' CMP e=2.6' nb=0.074

Barrel Shape: Circular

Barrel Diameter: 7.50 ft

Barrel Material: Corrugated Steel

Embedment: 31.20 in

Barrel Manning's n: 0.0240 (top and sides)

Manning's n: 0.0740 (bottom)

Inlet Type: Conventional

Inlet Edge Condition: Square Edge with Headwall

Inlet Depression: None

Table 2 - Downstream Channel Rating Curve (Crossing: Trib to Bear Crk n=0.074)

Flow (cfs)	Water Surface Elev (ft)	Depth (ft)	Velocity (ft/s)	Shear (psf)	Froude Number
0.00	313.50	0.00	0.00	0.00	0.00
2.40	313.88	0.38	2.05	2.00	0.79
4.80	313.99	0.49	2.64	2.61	0.83
7.20	314.08	0.58	3.05	3.10	0.85
9.60	314.17	0.67	3.36	3.53	0.87
12.00	314.24	0.74	3.63	3.92	0.88
14.40	314.31	0.81	3.85	4.28	0.89
16.80	314.37	0.87	4.05	4.61	0.90
19.20	314.43	0.93	4.23	4.92	0.91
21.60	314.48	0.98	4.39	5.22	0.91
24.00	314.54	1.04	4.54	5.50	0.92

Tailwater Channel Data - Trib to Bear Crk n=0.074

Tailwater Channel Option: Irregular Channel

Roadway Data for Crossing: Trib to Bear Crk n=0.074

Roadway Profile Shape: Constant Roadway Elevation

Crest Length: 99.00 ft

Crest Elevation: 325.00 ft

Roadway Surface: Paved

Roadway Top Width: 26.00 ft

I.3.3. HEC-RAS Output for 7.5 ft CMP at QH.

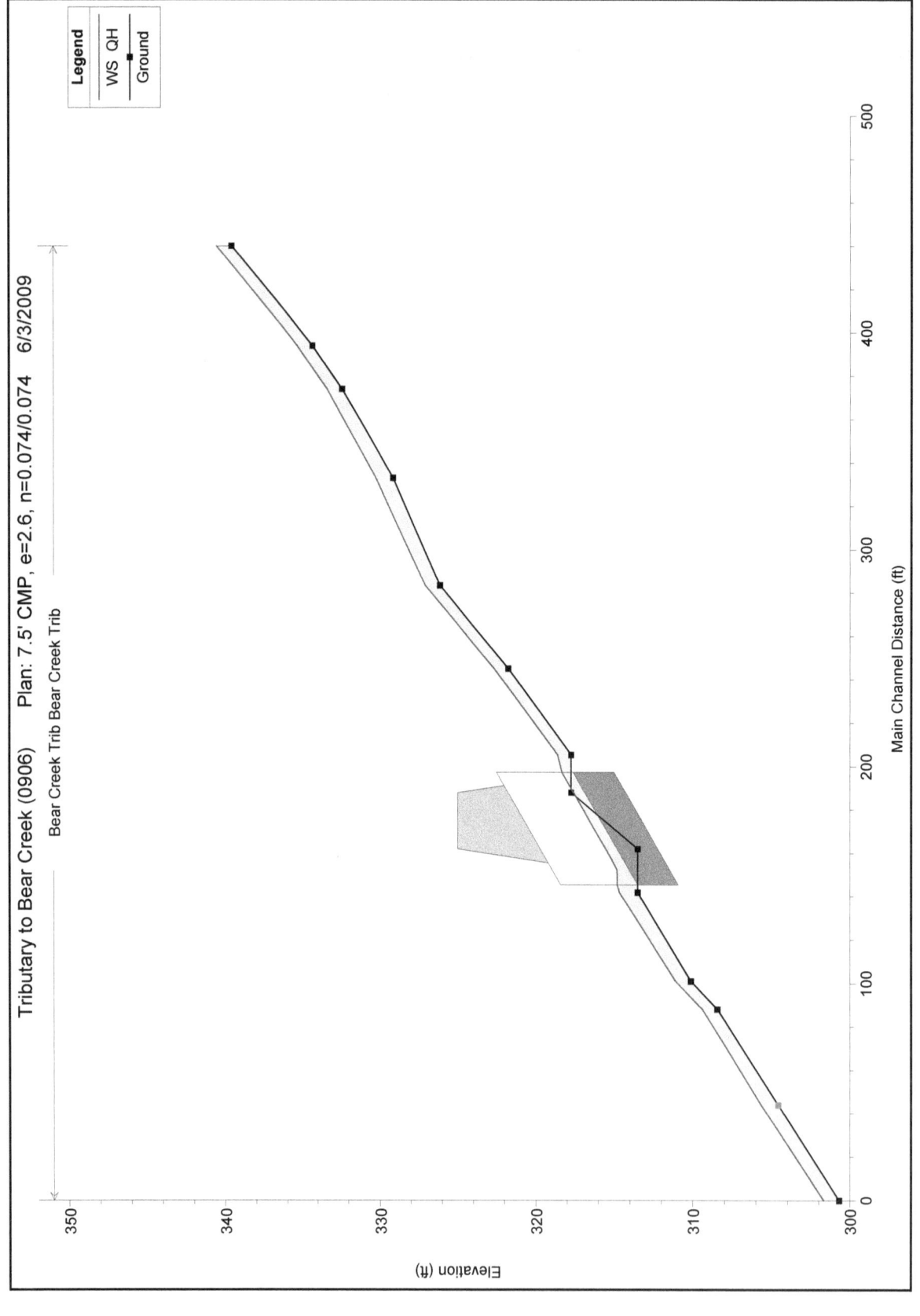

I.3.3. HEC-RAS Output for 7.5 ft CMP at QH.

HEC-RAS Plan: 7.5' 2.6' 0.074   River: Bear Creek Trib   Reach: Bear Creek Trib   Profile: QH

Reach	River Sta	Profile	Q Total (cfs)	Min Ch El (ft)	W.S. Elev (ft)	Max Chl Dpth (ft)	Hydr Depth (ft)	Mann Wtd Chnl	Crit W.S. (ft)	E.G. Elev (ft)	E.G. Slope (ft/ft)	Vel Chnl (ft/s)	Flow Area (sq ft)	Top Width (ft)	Froude # Chl
Bear Creek Trib	440	QH	24.00	339.57	340.54	0.97	0.60	0.078	340.54	340.84	0.121835	4.42	5.43	9.10	1.01
Bear Creek Trib	394	QH	24.00	334.31	335.34	1.03	0.63	0.074	335.28	335.59	0.082525	3.99	6.01	9.49	0.88
Bear Creek Trib	374	QH	24.00	332.39	333.35	0.96	0.66	0.078	333.35	333.68	0.109825	4.62	5.19	7.84	1.00
Bear Creek Trib	333	QH	24.00	329.13	330.25	1.12	0.83	0.070	330.05	330.44	0.040930	3.51	6.84	8.22	0.68
Bear Creek Trib	283.5	QH	24.00	326.13	327.07	0.94	0.67	0.077	327.05	327.38	0.103833	4.46	5.38	7.99	0.96
Bear Creek Trib	245.2	QH	24.00	321.75	322.67	0.92	0.65	0.080	322.67	323.00	0.126745	4.64	5.17	7.96	1.02
Bear Creek Trib	205.4	QH	24.00	317.73	318.60	0.87	0.62	0.058	318.65	318.99	0.082084	5.01	4.79	7.90	1.13
Bear Creek Trib	175		Culvert												
Bear Creek Trib	142	QH	24.00	313.50	314.65	1.15	0.85	0.080	314.49	314.89	0.065277	3.93	6.11	7.20	0.75
Bear Creek Trib	101	QH	24.00	310.10	311.09	0.99	0.72	0.077	311.09	311.45	0.110901	4.86	4.94	6.84	1.01
Bear Creek Trib	88	QH	24.00	308.44	309.37	0.93	0.68	0.077	309.43	309.81	0.143882	5.28	4.54	6.69	1.13
Bear Creek Trib	44.*	QH	24.00	304.57	305.65	1.08	0.83	0.074	305.55	305.94	0.070611	4.30	5.58	6.75	0.83
Bear Creek Trib	0	QH	24.00	300.69	301.68	0.99	0.73	0.077	301.68	302.05	0.113664	4.90	4.90	6.71	1.01

I.3.3. HEC-RAS Output for 7.5 ft CMP at QH.

Plan: 7.5' 2.6' 0.074   Bear Creek Trib   Bear Creek Trib RS: 175 Culv Group: Culvert #1   Profile: QH

Q Culv Group (cfs)	24.00	Culv Full Len (ft)	
# Barrels	1	Culv Vel US (ft/s)	4.51
Q Barrel (cfs)	24.00	Culv Vel DS (ft/s)	2.50
E.G. US. (ft)	318.99	Culv Inv El Up (ft)	315.00
W.S. US. (ft)	318.60	Culv Inv El Dn (ft)	310.90
E.G. DS (ft)	314.89	Culv Frctn Ls (ft)	3.75
W.S. DS (ft)	314.65	Culv Exit Loss (ft)	0.00
Delta EG (ft)	4.10	Culv Entr Loss (ft)	0.28
Delta WS (ft)	3.95	Q Weir (cfs)	
E.G. IC (ft)	318.45	Weir Sta Lft (ft)	
E.G. OC (ft)	318.93	Weir Sta Rgt (ft)	
Culvert Control	Outlet	Weir Submerg	
Culv WS Inlet (ft)	318.33	Weir Max Depth (ft)	
Culv WS Outlet (ft)	314.80	Weir Avg Depth (ft)	
Culv Nml Depth (ft)	3.33	Weir Flow Area (sq ft)	
Culv Crt Depth (ft)	3.30	Min El Weir Flow (ft)	325.01

Errors Warnings and Notes

Warning:	During subcritical analysis, the water surface upstream of culvert went to critical depth.
Note:	During the supercritical calculations a hydraulic jump occurred at the inlet of (going into) the
	culvert.

Table 1 - Culvert Summary Table: 12' CMP e=4.43' nb=0.074

Total Discharge (cfs)	Culvert Discharge (cfs)	Headwater Elevation (ft)	Inlet Control Depth (ft)	Outlet Control Depth (ft)	Flow Type	Normal Depth (ft)	Critical Depth (ft)	Outlet Depth (ft)	Tailwater Depth (ft)	Outlet Velocity (ft/s)	Tailwater Velocity (ft/s)
0.00	0.00	317.60	0.000	-4.100	0-NF	0.000	0.000	0.000	0.000	0.000	0.000
2.40	2.40	317.73	0.069	0.127	3-M1t	0.047	0.042	0.377	0.377	0.556	2.053
4.80	4.80	317.80	0.136	0.205	3-M1t	0.094	0.084	0.492	0.492	0.847	2.642
7.20	7.20	317.87	0.201	0.269	3-M1t	0.141	0.127	0.585	0.585	1.064	3.048
9.60	9.60	317.94	0.265	0.337	3-M1t	0.188	0.169	0.666	0.666	1.242	3.365
12.00	12.00	318.00	0.327	0.399	3-M1t	0.235	0.211	0.739	0.739	1.397	3.628
14.40	14.40	318.06	0.388	0.456	3-M1t	0.283	0.253	0.806	0.806	1.534	3.854
16.80	16.80	318.12	0.449	0.520	3-M1t	0.330	0.295	0.869	0.869	1.658	4.053
19.20	19.20	318.18	0.508	0.581	3-M1t	0.377	0.337	0.928	0.928	1.772	4.232
21.60	21.60	318.24	0.567	0.642	3-M1t	0.424	0.380	0.984	0.984	1.878	4.394
24.00	24.00	318.31	0.627	0.706	3-M1t	0.471	0.422	1.037	1.037	1.977	4.543

I.3.4. HY8 Report for 12.0 ft CMP with Oversize Bed at QH.

****

Inlet Elevation (invert): 317.60 ft,   Outlet Elevation (invert): 313.50 ft

Culvert Length: 52.06 ft,   Culvert Slope: 0.0790

****

Site Data - 12' CMP e=4.43' nb=0.074

Site Data Option: Culvert Invert Data

Inlet Station: 0.00 ft

Inlet Elevation: 313.17 ft

Outlet Station: 51.90 ft

Outlet Elevation: 309.07 ft

Number of Barrels: 1

Culvert Data Summary - 12' CMP e=4.43' nb=0.074

Barrel Shape: Circular

Barrel Diameter: 12.00 ft

Barrel Material: Corrugated Steel

Embedment: 53.16 in

Barrel Manning's n: 0.0240 (top and sides)

Manning's n: 0.0740 (bottom)

Inlet Type: Conventional

Inlet Edge Condition: Square Edge with Headwall

Inlet Depression: None

I.3.4. HY8 Report for 12.0 ft CMP with Oversize Bed at QH.

Table 2 - Downstream Channel Rating Curve (Crossing: Trib to Bear Crk n=0.074)

Flow (cfs)	Water Surface Elev (ft)	Depth (ft)	Velocity (ft/s)	Shear (psf)	Froude Number
0.00	313.50	0.00	0.00	0.00	0.00
2.40	313.88	0.38	2.05	2.00	0.79
4.80	313.99	0.49	2.64	2.61	0.83
7.20	314.08	0.58	3.05	3.10	0.85
9.60	314.17	0.67	3.36	3.53	0.87
12.00	314.24	0.74	3.63	3.92	0.88
14.40	314.31	0.81	3.85	4.28	0.89
16.80	314.37	0.87	4.05	4.61	0.90
19.20	314.43	0.93	4.23	4.92	0.91
21.60	314.48	0.98	4.39	5.22	0.91
24.00	314.54	1.04	4.54	5.50	0.92

Tailwater Channel Data - Trib to Bear Crk n=0.074

Tailwater Channel Option: Irregular Channel

Roadway Data for Crossing: Trib to Bear Crk n=0.074

Roadway Profile Shape: Constant Roadway Elevation

Crest Length: 99.00 ft

Crest Elevation: 325.00 ft

Roadway Surface: Paved

Roadway Top Width: 26.00 ft

I.3.5. HEC-RAS Output for 12.0 ft CMP with Oversized Bed at QH.

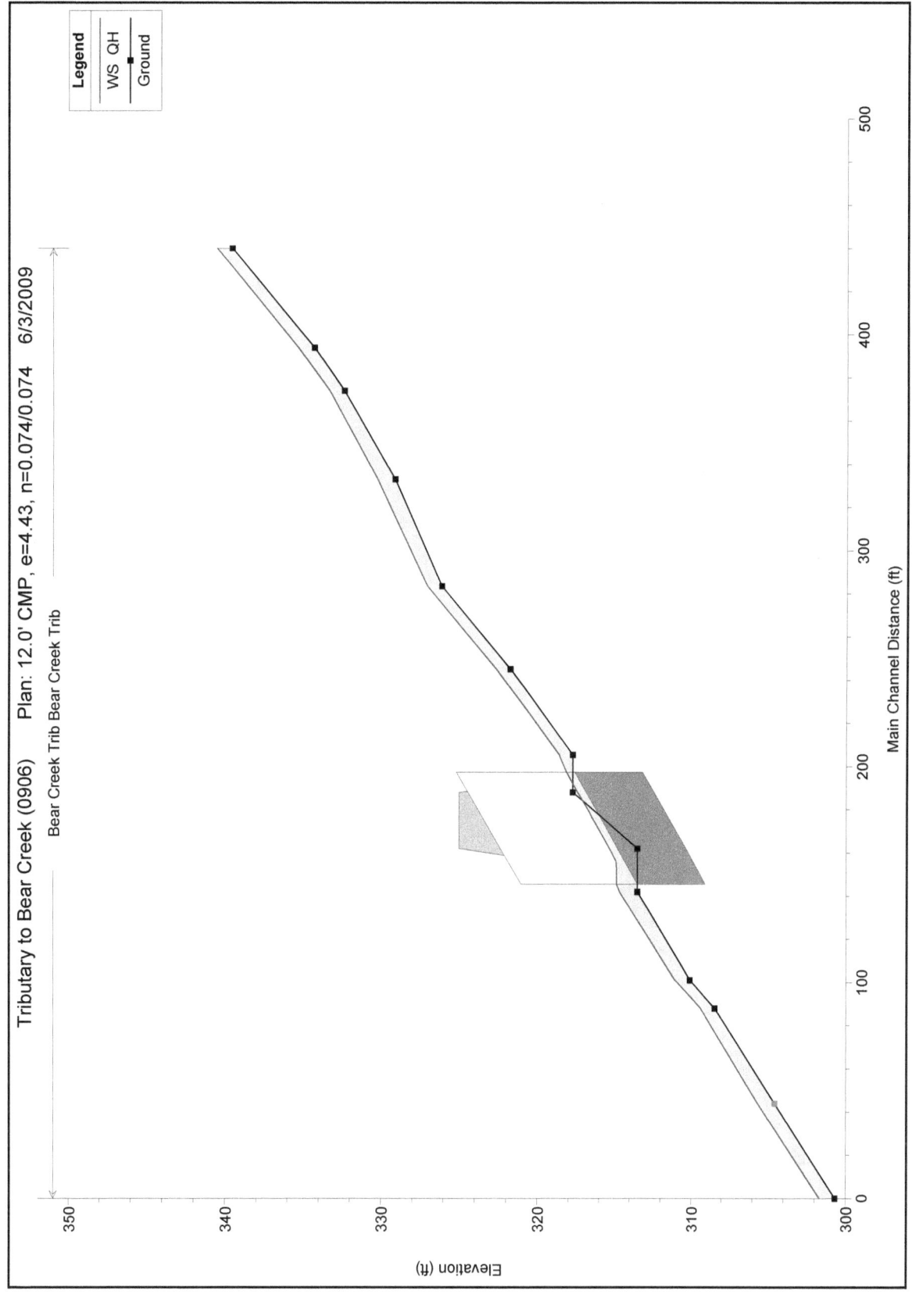

Tributary to Bear Creek (0906)   Plan: 12.0' CMP, e=4.43, n=0.074/0.074   6/3/2009

Bear Creek Trib Bear Creek Trib

Legend

WS QH
Ground

I.3.5. HEC-RAS Output for 12.0 ft CMP with Oversized Bed at QH.

HEC-RAS Plan: 12.0 4.43 0.074   River: Bear Creek Trib   Reach: Bear Creek Trib   Profile: QH

Reach	River Sta	Profile	Q Total (cfs)	Min Ch El (ft)	W.S. Elev (ft)	Max Chl Dpth (ft)	Hydr Depth (ft)	Mann Wtd Chnl	Crit W.S. (ft)	E.G. Elev (ft)	E.G. Slope (ft/ft)	Vel Chnl (ft/s)	Flow Area (sq ft)	Top Width (ft)	Froude # Chl
Bear Creek Trib	440	QH	24.00	339.57	340.54	0.97	0.60	0.078	340.54	340.84	0.121835	4.42	5.43	9.10	1.01
Bear Creek Trib	394	QH	24.00	334.31	335.34	1.03	0.63	0.074	335.28	335.59	0.082525	3.99	6.01	9.49	0.88
Bear Creek Trib	374	QH	24.00	332.39	333.35	0.96	0.66	0.078	333.35	333.68	0.109825	4.62	5.19	7.84	1.00
Bear Creek Trib	333	QH	24.00	329.13	330.25	1.12	0.83	0.070	330.05	330.44	0.040930	3.51	6.84	8.22	0.68
Bear Creek Trib	283.5	QH	24.00	326.13	327.07	0.94	0.67	0.077	327.05	327.38	0.103833	4.46	5.38	7.99	0.96
Bear Creek Trib	245.2	QH	24.00	321.75	322.67	0.92	0.65	0.080	322.67	323.00	0.126745	4.64	5.17	7.96	1.02
Bear Creek Trib	205.4	QH	24.00	317.73	318.60	0.87	0.61	0.058	318.66	318.99	0.081772	4.96	4.84	7.91	1.12
Bear Creek Trib	175		Culvert												
Bear Creek Trib	142	QH	24.00	313.50	314.65	1.15	0.85	0.080	314.49	314.89	0.065277	3.93	6.11	7.20	0.75
Bear Creek Trib	101	QH	24.00	310.10	311.09	0.99	0.72	0.077	311.09	311.45	0.110901	4.86	4.94	6.84	1.01
Bear Creek Trib	88	QH	24.00	308.44	309.37	0.93	0.68	0.077	309.43	309.81	0.143882	5.28	4.54	6.69	1.13
Bear Creek Trib	44.*	QH	24.00	304.57	305.65	1.08	0.83	0.074	305.55	305.94	0.070611	4.30	5.58	6.75	0.83
Bear Creek Trib	0	QH	24.00	300.69	301.68	0.99	0.73	0.077	301.68	302.05	0.113664	4.90	4.90	6.71	1.01

I.3.5. HEC-RAS Output for 12.0 ft CMP with Oversized Bed at QH.

Plan: 12.0 4.43 0.074   Bear Creek Trib   Bear Creek Trib   RS: 175   Culv Group: Culvert #1   Profile: QH

Q Culv Group (cfs)	24.00	Culv Full Len (ft)	
# Barrels	1	Culv Vel US (ft/s)	3.75
Q Barrel (cfs)	24.00	Culv Vel DS (ft/s)	1.49
E.G. US. (ft)	318.99	Culv Inv El Up (ft)	313.17
W.S. US. (ft)	318.60	Culv Inv El Dn (ft)	309.07
E.G. DS (ft)	314.89	Culv Frctn Ls (ft)	3.47
W.S. DS (ft)	314.65	Culv Exit Loss (ft)	0.00
Delta EG (ft)	4.09	Culv Entr Loss (ft)	0.20
Delta WS (ft)	3.95	Q Weir (cfs)	
E.G. IC (ft)	318.06	Weir Sta Lft (ft)	
E.G. OC (ft)	318.56	Weir Sta Rgt (ft)	
Culvert Control	Outlet	Weir Submerg	
Culv WS Inlet (ft)	318.15	Weir Max Depth (ft)	
Culv WS Outlet (ft)	314.86	Weir Avg Depth (ft)	
Culv Nml Depth (ft)	4.98	Weir Flow Area (sq ft)	
Culv Crt Depth (ft)	4.94	Min El Weir Flow (ft)	325.01

Errors Warnings and Notes

Warning:	During subcritical analysis, the water surface upstream of culvert went to critical depth.
Note:	During the supercritical calculations a hydraulic jump occurred at the inlet of (going into) the
	culvert.

I.4 REFERENCES

Alaska Department of Transportation and Public Facilities, 1995. "Alaska Highway Drainage Manual," June.

Curran, Janet H., David F. Meyer, and Gary D. Tasker, 2003. "Estimating the Magnitude and Frequency of Peak Streamflows for Ungaged Sites on Streams in Alaska and Conterminous Basins in Canada," USGS Water-Resources Investigations Report 03-4188.

Gubernick, Robert, 1995. "Tongass National Forest Fish Passage Flow Determination Methods," unpublished document.

Wiley, Jeffrey B. and Janet H. Curran, 2003. "Estimating Annual High-Flow Statistics and Monthly and Seasonal Low-Flow Statistics for Ungaged Sites on Streams in Alaska and Conterminous Basins in Canada," USGS Water-Resources Investigations Report 03-4114.

APPENDIX J- DESIGN EXAMPLE: SICKLE CREEK, MICHIGAN

J.1 SITE DESCRIPTION

The design procedure is applied to a road crossing of Sickle Creek, which is approximately 3 miles (4.8 kilometers) south of Norwalk, Michigan. The drainage area to the crossing is 3.2 mi^2 (8.3 km^2). (Data and photos for this application were taken from the FishXing web site and provided by Stephanie Ogren of the Little River Band of Ottawa Indians.)

Prior to their replacement, twin 36-in (910-mm) CMP culverts were operating at the stream-road crossing. Both of the 36-ft (11-m) long culverts were perched 1 ft (0.3 m) above the downstream scour pool creating a jump barrier. Assessment also indicated that the culverts constricted the channel creating excessive velocities. Figure J.1 shows the outlet of the culvert.

Figure J.1.   Sickle Creek Outlet.

J.2 DESIGN PROCEDURE APPLICATION

This section illustrates the application of the design procedure. The uses of two separate tool sets are shown: 1) HY-8 with normal depth computations for the channel cross-sections and 2) HEC-RAS. Although both tool sets are shown, the designer may choose one or the other as appropriate for the site and the designers modeling skills.

Step 1. Determine Design Flows.

Discharges are determined for the peak flow, Q_P, high passage flow, Q_H, and low passage flow, Q_L. For the peak flow, Q_p, the Michigan Department of Transportation standards for culvert design are applied for this site. The applicable criterion for culverts is a 50-yr event. Selected watershed and rainfall characteristics used to develop the design flows are summarized in Table J.1.

Holtschlag and Croskey (1984) provide a set of USGS peak flow regression equations applicable to this site for the 5-yr to 100-yr events. The 50-yr design event for Q_P is shown in Table J.2 as 163 ft^3/s 4.62 m^3/s). A second methodology available is provided by Sorrell (2008). This method is a variation on the NRCS curve number method. It results in a 50-yr discharge estimate nearly twice as high as the regression equations. On the FishXing web site, the 100-yr discharge is reported for this watershed as 230 ft^3/s 6.51 m^3/s). The method used to develop this estimate was not provided. Reviewing the 50-yr and 100-yr estimates in Table J.2, it is apparent that the Sorrell method results in discharges higher than the other two sources, which are somewhat consistent with each other. Therefore, a discharge of 163 ft^3/s 4.62 m^3/s) will be used for Q_p.

Table J.1.   Watershed and Rainfall Characteristics.

Characteristic	CU	SI
Drainage area	3.2 mi^2	8.3 km^2
Main channel slope	62 ft/mi	11.7 m/km
100-yr 24-h rainfall	5.08 in	129 mm
2-yr 24-h rainfall	2.09 in	53 mm
Curve number	59	59
Time of concentration	2.9 h	2.9 h

Table J.2.   Discharge Estimates.

Discharge Quantity	Holtschlag and Croskey (1984) Region 3, ft^3/s (m^3/s)	Sorrell (2008), ft^3/s (m^3/s)	FishXing Website, ft^3/s (m^3/s)
Q_{100}	188 (5.32)	410 (11.6)	230 (6.51)
Q_{50}	Q_P = 163 (4.62)	300 (8.50)	
Q_2	60 (1.70)	20 (0.57)	
$Q_{10\%}$	3.6 (0.10)		
$0.25Q_2$	Q_H = 15 (0.42)	4 (0.14)	
$Q_{75\%}$	1.8 (0.051)		
$Q_{95\%}$	1.5 (0.042)		
7Q2	Q_L = 1.6 (0.045)		

The high passage flow is determined by site-specific guidelines, if they exist. None are known to exist. In the absence of site-specific guidelines, the Q_H may be defined as the 10 percent exceedance quantile on the annual flow duration curve. A flow duration curve does not exist for this location, but Holtschlag and Croskey (1984) include a regression equation that results in a 10 percent exceedance flow based on daily data. The 10 percent exceedance flow may also be estimated as 25 percent of the Q_2.

Table J.2 summarizes the high passage flow estimates available ranging from 15 ft^3/s (0.42 m^3/s) (25 percent of the Holtschlag and Croskey Q_2) and 3.6 ft^3/s (0.10 m^3/s) (Holtschlag and Croskey 10 percent exceedance). Given the wide range of estimates, the high passage flow will be taken to be 15 ft^3/s (0.42 m^3/s) to be conservative in characterizing the range of passage flows for design.

The low passage flow is determined by site-specific guidelines, if they exist. None are known to exist for this site. In the absence of site-specific guidelines, the Q_L should be defined as the 90 percent exceedance quantile on the annual flow duration curve or the 7-day, 2-yr low flow (7Q2), but no less than 1 ft^3/s (0.028 m^3/s). As previously noted, a flow duration curve does not exist for this location, but Holtschlag and Croskey (1984) include regression equations for both the 95 percent and 75 percent exceedance flows based on daily data, as well as the 7Q2.

These values are summarized in Table J.2. These estimates are relatively consistent with one another. Q_L will be taken as 1.6 ft³/s (0.045 m³/s).

Step 2. Determine Project Reach and Representative Channel Characteristics.

The project reach should extend no less than three culvert lengths or 200 ft (61 m), whichever is greater, up and downstream of the crossing location. Since the existing culvert is 36 ft (11 m) in length, the project reach must extend at least 200 ft (61 m) upstream and downstream of the culvert inlet and outlet, respectively. At least three cross-sections should be obtained upstream and downstream from the crossing location.

Project reach data were derived from DeBoer, et al. (2007) and Ogren, et al. (2008) as well as from digital elevation data from the USGS. Although preconstruction surveyed cross-sections were not available, eight cross sections were composited from the three sources previously mentioned and a partial field survey in 2009. Four are downstream of the culvert; the most downstream cross-section is approximately 207 ft (63.1 m) downstream from the existing culvert outlet.

Four cross-sections are upstream; the most upstream cross-section is approximately 162 ft (49.3 m) upstream of the existing culvert inlet. Although, this cross-section location does not achieve the full 200 ft (61 m) criterion, the distance is over 4 culvert lengths and is, therefore, judged to be sufficient. Table J.3 summarizes the cross-section locations and Figure J.2 shows the creek and cross-sections schematically in plan view. Cross-sections shown in Figure J.2, but not listed in Table J.3 were interpolated/extrapolated for the purpose of water surface profile modeling with HEC-RAS. Plots of the cross-sections are included in Section J.3.1.

Table J.3. Composited Cross-sections.

Cross-section	Station (ft)	Station (m)	Thalweg Elevation (ft)	Thalweg Elevation (m)	Slope to downstream cross-section (ft/ft or m/m)
875	875	267	289.54	88.25	0.0053
823	823	251	289.26	88.17	0.0054
772	772	235	288.99	88.08	0.0133
719	719	219	288.29	87.87	0.0061
Road centerline	695	212	--	--	--
666	666	203	287.97	87.77	0.0084
614	614	187	287.53	87.64	0.0079
550	550	168	287.02	87.48	0.0056
470	470	143	286.57	87.35	0.0082

The longitudinal profile of the stream and existing roadway embankment is shown in Figure J.3 using a composited thalweg. Superimposed on the detailed profile are the cross-section locations plotted with their thalweg elevations as well as the existing culvert. The existing culvert slope is 1.4 percent. A local scour hole is noted at the outlet and a small depression at the inlet.

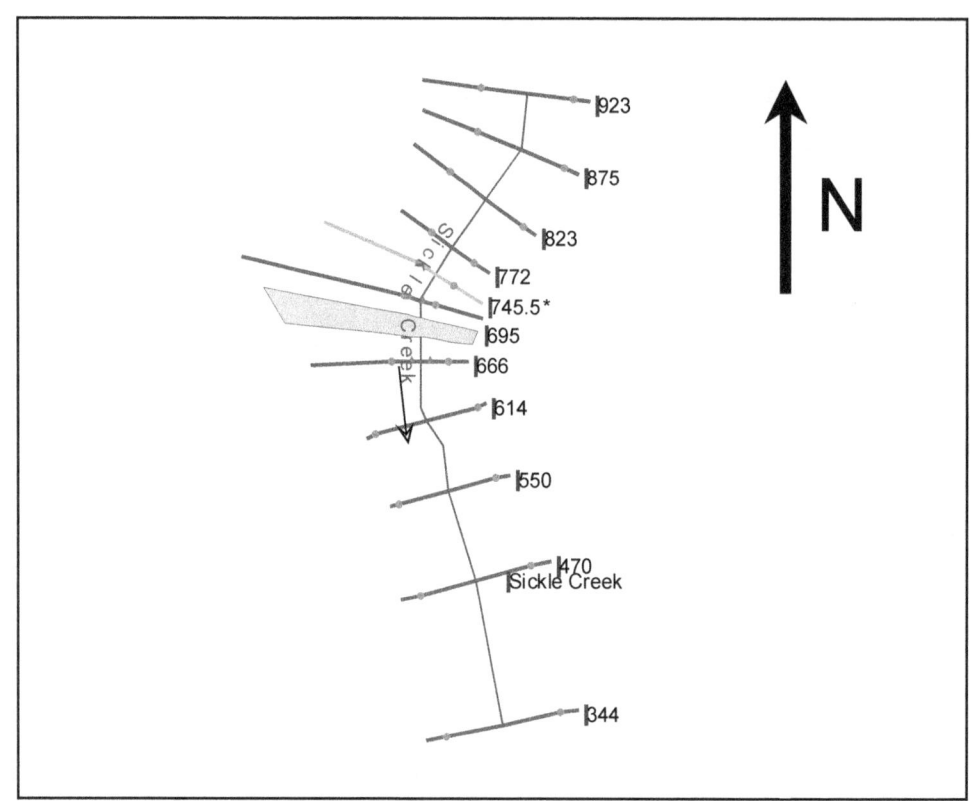

Figure J.2. Creek and Cross-section Schematic.

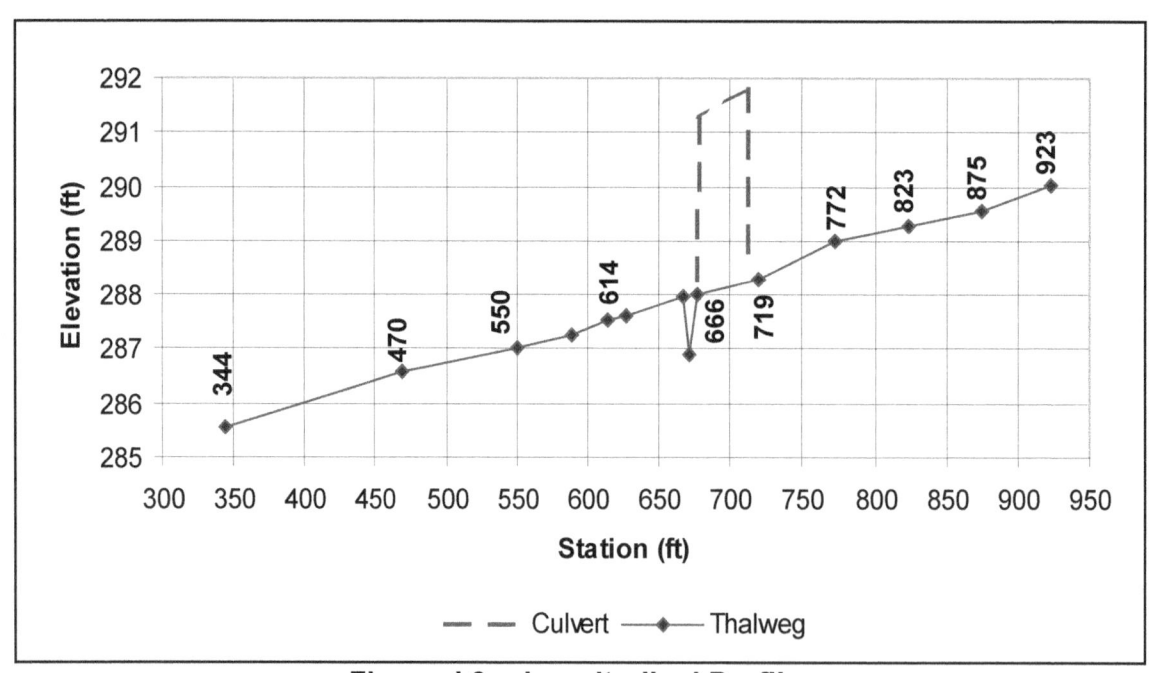

Figure J.3. Longitudinal Profile.

Several bed material samples were taken upstream and downstream of the road annually from 2004 through 2007 (DeBoer, et al., 2007). Since the creek bed is sandy, it is mobile. However, the gradation over time and up and downstream of the road is reasonably consistent. The average gradation from these samples is summarized in Figure J.4 along with the maximum and minimum values. Table J.4 summarizes the average gradation. Evidence of bed armoring was not reported. The unit weight of the bed material was not provided so a value of 156 lb/ft³ (24,500 N/m³) was assumed.

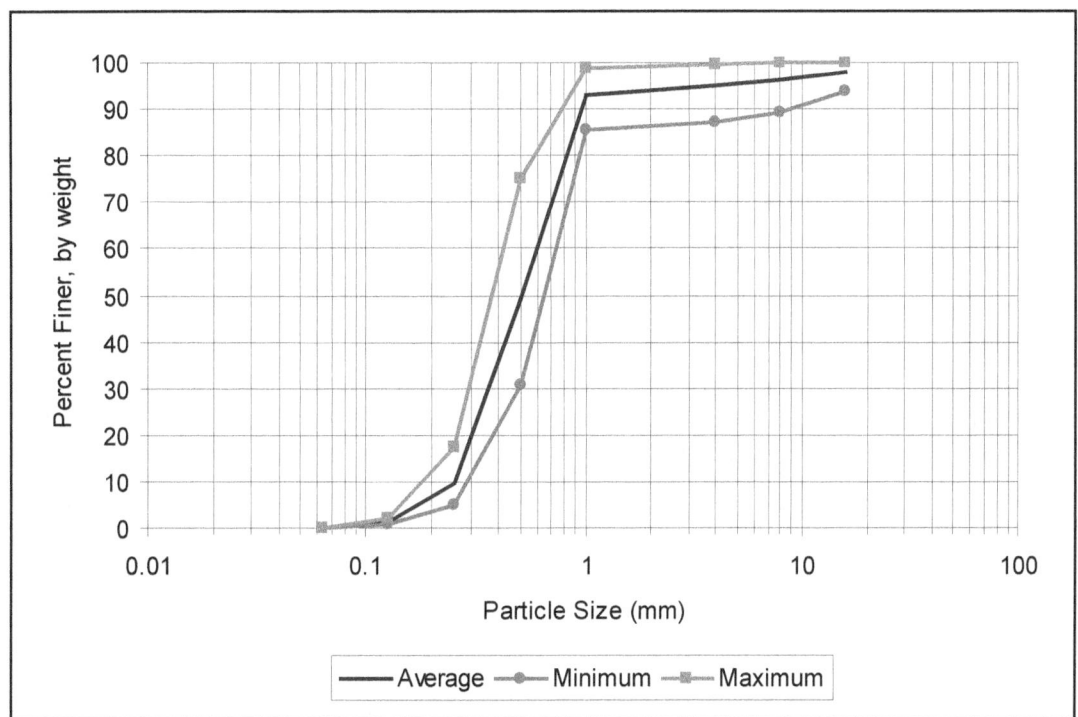

Figure J.4.   Measured Bed Material Gradations.

Table J.4.   Average Bed Material Quantiles.

Quantile	Size (in)	Size (ft)	Size (mm)
D_{95}	0.140	0.0117	3.55
D_{84}	0.034	0.0028	0.87
D_{50}	0.020	0.0017	0.50
D_{16}	0.011	0.00092	0.28
D_5	0.007	0.00058	0.17

Step 3. Check for Dynamic Equilibrium.

The qualitative assessment for dynamic equilibrium involves three components:

1. Watershed reconnaissance for changes in supply.

2. Project reach sediment transport assessment.

3. Field observations of the project reach.

The watershed reconnaissance is to identify changes in the watershed that may result in changes in sediment supply. Activities in the watershed include recreational activities, such as fishing.

The project reach sediment transport assessment examines potential changes in discharge, slope, and D_{50} throughout the project reach in the context of Lane's proportional relationship to sediment transport (Equation 7.2). Discharge throughout the reach is invariant and the sediment size appears relatively homogeneous. Average slopes in the project reach range from 0.005 ft/ft to 0.013 ft/ft (see Table J.3). Slope variations are not considered to be an issue for sediment transport.

Field observations are to identify any indicators of instability. No instabilities were reported.

Taking the components together, there is no clear concern that the project reach is experiencing instability or disequilibrium in sediment transport. The design for the passage culvert may improve sediment transport equilibrium.

Step 4. Analyze and Mitigate Channel Instability.

Based on the assessment in Step 3, this step is unnecessary.

Step 5. Align and Size Culvert for Q_p.

Criteria allow for a headwater depth to culvert rise ratio (HW/D ratio) of up to 1.2 provided the headwater does not damage upstream property and the flow is not diverted away from the culvert. An embedded CMP culvert will be designed with these criteria.

The existing CMP culvert pair (two 3-ft diameter pipes on a slope of 1.4 percent) analyzed first. For the peak design flow, Q_P, of 163 cfs the existing culvert (without embedment) has a HW/D ratio of 1.6 and overtops the road according to an analysis using HY-8. Therefore, the existing culvert does not meet hydraulic criteria.

The horizontal alignment of the existing culvert will be maintained.

The desired vertical alignment of the replacement culvert is established by evaluation of the vertical profile of the stream and the existing culvert profile. The profile in Figure J.3 averages 0.007 ft/ft in slope. For the initial design trial, the culvert will be laid out as close as practicable to the average project reach slope.

An initial CMP culvert diameter of 7.0 ft is estimated considering that two 3-ft diameter culverts (existing) are inadequate and a 2 ft minimum embedment is required. The embedment criteria for a circular culvert is 30 percent of the culvert rise giving an embedded depth of 0.3 x 7.0 ft = 2.1 ft. In addition, embedment depth may be no less than 2.0 ft or 2 times the D_{95} of the bed material. Since the bed material is sand, the D_{95} is not limiting. Therefore, the minimum embedment depth is 2.1 ft. Inlet and outlet elevations, for the existing and replacement culvert are summarized in Table J.5. With these elevations, the proposed culvert slope is 0.007, which approximates the average reach slope. Within construction tolerances, the 0.25 ft drop over the 36 ft culvert length is close to a zero slope.

Table J.5.   Inlet and Outlet Elevations for Existing and Replacement Culverts.

Description	Inlet	Outlet
Existing Culvert Invert	288.8	288.3
Replacement Culvert Bed	288.25	288.0
Replacement Culvert Invert	286.15	285.9

A bed gradation must be selected. Since we have several samples, the average gradation sample from the site will be selected for design as is shown in Table J.6. Recall that to control bed interstitial flow, it is recommended that the D_5 fraction be no larger than 2 mm (sand, silt, and clay). The existing gradation satisfies this requirement; no modification is required.

Table J.6.   Bed Gradation Design.

Quantile	Design (mm)	Design (ft)	Design (in)
D_{95}	3.55	0.01167	0.140
D_{84}	0.87	0.00283	0.034
D_{50}	0.50	0.00167	0.020
D_{16}	0.28	0.00092	0.011
D_5	0.17	0.00058	0.007

A Manning's n is needed to estimate the roughness of the bed material in the culvert. Richardson, et al. (2001) provide a description of the roughness behavior of sand bed channels. A simplified procedure is found in Appendix C. At Q_P, an assessment of the stream power in the channel and the grain size results in the conclusion that the channel is experiencing the upper flow regime. Based on the range of Manning's n appropriate for this condition, n = 0.016 is selected. Although, lower stream power is computed at Q_H, sufficient stream power is generated to experience the upper flow regime. n = 0.016 is also selected for this condition. At Q_L, assessment of the stream power results in the conclusion that the lower flow regime occurs with a dune bedform. This bedform exhibits a higher roughness value leading to the selection of n = 0.028. Manning's roughness values are summarized in Table J.7.

Table J.7.   Manning's n for Sand Bed Channel.

Discharge (cfs)	Manning's n
1.6 (Q_L)	0.028
15 (Q_H)	0.016
163 (Q_P)	0.016

The Manning's n corresponding with Q_P is used for the bed material in this step. For these conditions the culvert operates in outlet control with a headwater depth of 4.6 ft. The HW/D ratio is 4.6/(7.0-2.1) = 0.9, which is less than or equal to the 1.2 maximum criteria. Since the criterion is satisfied by a large margin we could evaluate a smaller barrel. However, we will see in later steps that the 7-ft size is not sufficient.

The headwater also does not overtop the road or result in a redirection of flows away from the culvert. Although this culvert has a higher rise than the existing culvert, minimum cover still appears to be available. Proceed to Step 6.

Step 6. Check Culvert Bed Stability at Q_H

Using a Manning's n of 0.016, the shear stress in the culvert is compared with the permissible shear stress for the bed material. The permissible shear stress for fine-grained, non-cohesive soils (D_{75} < 1.3 mm (0.05 in)) is relatively constant and is conservatively estimated to be 1.0 N/m^2 (0.02 lb/ft^2) (Kilgore and Cotton, 2005).

Applied shear stress is estimated at the inlet and outlet of the culvert using Equation 7.9:

$\tau_d = \gamma y S$

According to HY-8, the culvert is in inlet control with a flow type 1-S2n. From the HY-8 report, these depths are 0.54 ft and 0.53 ft, respectively.

The energy slope, S, is estimated from Manning's equation once velocity and composite n are computed. The wetted perimeter of the wall is based on the circular arc.

For the inlet, the calculations are:

$y = 0.54$ ft

$A = 3.573$ ft^2

$P_{bed} = 6.416$ ft

$P_{wall} = 1.143$ ft

$P_{total} = 7.558$ ft

$R_h = 3.573/7.558 = 0.473$ ft

$$n_{comp} = \left[\frac{P_{bed}n_{bed}^{1.5} + P_{wall}n_{wall}^{1.5}}{P_{bed} + P_{wall}}\right]^{2/3} = \left[\frac{6.416(0.016)^{1.5} + 1.143(0.024)^{1.5}}{7.558}\right]^{2/3} = 0.017$$

$V = Q/A = 15/3.573 = 4.20$ ft/s

$$S = \left(\frac{V(n)}{1.49\left(R_h^{2/3}\right)}\right)^2 = \left(\frac{4.20(0.017)}{1.49(0.473)^{2/3}}\right)^2 = 0.0062 \; ft/ft$$

$\tau_d = \gamma y S = 6.24(0.54)(0.0062) = 0.2$ lb/ft^2

For the outlet, the calculations are performed in the same way. The HY-8 results (see Section J.3.2) are summarized in Table J.8. Comparing inlet and the outlet conditions, the highest shear stress is at the outlet estimated as 0.4 lb/ft^2. Since this is well above the permissible shear stress of the sandy bed material, the culvert bed is not stable at Q_H.

Alternatively, using HEC-RAS (Section J.3.3), the depths at the culvert inlet and outlet are computed to be 0.55 and 0.46 ft, respectively, with the culvert operating under outlet control. The depths, velocities, maximum shear stresses, and permissible shear stress are summarized in Table J.8. Since the maximum shear stresses are well above the permissible shear stress, the culvert bed is not stable at Q_H.

Table J.8.   7.0-ft CMP Culvert Inlet and Outlet Parameters at Q_H.

Parameter*	HY-8		HEC-RAS	
	Inlet	Outlet	Inlet	Outlet
y (ft)	0.54	0.53	0.55	0.61
V (ft/s)	4.20	4.28	4.12	3.70
S_e (ft/ft)	0.0062	0.0066	0.0059	0.0042
τ_d (lbs/ft^2)	0.2	0.2	0.2	0.2
τ_p (lbs/ft^2)	0.02	0.02	0.02	0.02

*Embedment = 2.1 ft, n = 0.016.

Regardless of which tool was chosen, we conclude the culvert bed is not stable at Q_H and we continue with Step 7.

Step 7. Check Channel Bed Mobility at Q_H.

The assessment in Step 6 concluded that the bed material in the culvert bottom is not stable at Q_H. In this step, we evaluate whether material is moving at this discharge in the upstream and the downstream channels. Table J.9 summarizes shear stress estimates at the cross-sections upstream and downstream using alternative methods: 1) a normal depth assumption and 2) HEC-RAS. The table also provides the computed shear stresses at the inlet and outlet of the culvert and the permissible shear stress for the bed material.

Table J.9.   Estimated Shear Stresses at Q_H.

Cross-section**	Normal Depth/HY-8 (lb/ft^2)	HEC-RAS (lb/ft^2)
875	0.3	0.1
823	0.3	0.2
772	0.6	0.3
Culvert Inlet*	0.2	0.2
Culvert Outlet*	0.2	0.2
666	0.3	0.1
614	0.3	0.2
550	0.3	0.4
470	0.4	0.4
τ_p (lbs/ft^2)	0.02	0.02

*7.0 ft CMP, embedment = 2.1 ft, n = 0.016.
**n=0.016.

Although cross-section estimates vary between the alternative estimation methods, all cross-sections exhibit shear stresses in excess of the permissible shear stress of 0.02 lb/ft^2. As may be expected with a sand bed channel ($D_{84} < 2$ mm), the bed material is mobile. Furthermore, the values determined within the culvert are within the range of stresses experienced in the channel. Therefore, it is concluded that the bed material is mobile and the culvert has not altered the ability of the material to move through the culvert (based on the HEC-RAS analysis). Proceed to Step 8.

Step 8. Check Culvert Bed Stability at Q_P.

Since the bed material is not stable at Q_H, it will also not be stable at the higher Q_P discharge. Therefore, we will need to design a stable bed in Step 9.

Step 9. Design Stable Bed for Q_P.

A stable bed design is attempted to resist the shear stresses at Q_P within the culvert. The bed will consist of a top layer of native material and an oversized underlayer. Design of the underlayer assumes the native top layer has been washed away at or before the peak of the hydrograph. It is assumed that natural replenishment cannot be relied on to restore the bed material in the culvert. At Q_P, the shear stresses in the culvert are expected to be greater than those in the channel since the flow is confined. If, however, site-specific analysis to the contrary is performed, the stable sublayer may be avoided.

As a first trial in designing the sublayer, select an oversized bed material that fits within the current culvert embedment of 2.1 ft. In accordance with the embedment criteria for Step 9, we would provide a 1 ft layer of native material leaving 1.1 ft for the oversized bed material. For a

CMP culvert, the oversize layer minimum embedment is $1.5D_{95}$, therefore, D_{95} = 1.1 ft/1.5 = 0.73 ft. Using the relation in Equation 7.15c between D_{50} and D_{95} for an oversized bed, the D_{50} = D_{95}/1.9 = 0.73 ft/1.9 = 0.38 ft. However, we will learn that this bed will not be stable at Q_P.

After subsequent trials, we select an oversized bed material that fits within a 10 ft CMP with a total embedment of 3.0 ft (30 percent of the culvert rise). In accordance with the embedment criteria, we would provide a 1 ft layer of native material leaving 2.0 ft for the oversized bed material. For a CMP culvert, the oversize layer minimum embedment is $1.5D_{95}$, therefore, D_{95} = 2.0 ft/1.5 = 1.33 ft. Using the relation in Equation 7.15c between D_{50} and D_{95} for an oversized bed, the D_{50} = D_{95}/1.9 = 1.33 ft/1.9 = 0.70 ft.

Assuming the native layer is washed out, we use HY-8 iteratively to determine that the normal depth in the culvert is 3.93 ft with a Manning's n of 0.052 for the oversize layer from the Limerinos equation. From this, a Shield's parameter of 0.052 is determined. The permissible shear stress is calculated from Equation 7.16:

$$\tau_p = 1.1F_*(\gamma_s - \gamma)D_{50} = 1.1(0.052)(156 - 62.4)(0.70) = 3.7 \text{ lbs/ft}^2$$

Applied shear stress is estimated at the inlet and outlet of the culvert based on the estimated depths. The inlet and outlet depths are taken from the water surface profile data for Q_P in HY-8 and are 3.20 ft and 2.37 ft, respectively (see section J.3.4). The resulting shear stresses are summarized in Table J.10. The applied shear stress is less than the permissible shear stress, therefore the bed is stable.

Alternatively, if HEC-RAS is used to analyze the culvert, the depths at the culvert inlet and outlet are computed to be 3.51 and 3.54 ft, respectively (see section J.3.5). The resulting shear stresses are summarized in Table J.10. The applied shear stress is less than the permissible shear stress, therefore the bed is stable.

Table J.10. 10.0 ft CMP Culvert Inlet and Outlet Parameters at Q_P.

Parameter*	HY-8		HEC-RAS	
	Inlet	Outlet	Inlet	Outlet
y (ft)	3.20	2.37	3.51	3.54
V (ft/s)	5.42	7.486	4.91	4.87
S_e (ft/ft)	0.0082	0.0227	0.0063	0.0061
τ_d (lbs/ft^2)	1.7	3.4	1.4	1.4
τ_p (lbs/ft^2)	3.7	3.7	3.7	3.7

*Embedment=2.0 ft (native layer washed out), S_o=0.007 ft/ft, n_{bed}=0.052.

Regardless of the method selected, the culvert bed is stable at Q_P, therefore, we continue by completing the oversized bed gradation design.

D_{84} is computed from Equation 7.15b:

$$D_{84} = 1.4D_{50} = 1.4(0.70) = 0.98 \text{ ft}$$

The D_5 is taken to be no larger than 2 mm to limit interstitial flow. The D_{16} is selected to provide a transition between the D_{50} and D_5 sizes. A reasonable transition is determined graphically. Table J.11 summarizes the resulting gradation and compares it to the native bed gradation.

Table J.11. Oversize Stable Bed Design Gradation.

Quantile	Native (mm)	Oversize (mm)	Native (ft)	Oversize (ft)
D_{95}	3.55	405	0.01167	1.33
D_{84}	0.87	299	0.00283	0.98
D_{50}	0.50	213	0.00167	0.70
D_{16}	0.28	21	0.00092	0.069
D_5	0.17	2	0.00058	0.0066

Proceed to Step 10.

Step 10. Check Culvert Velocity at Q_H.

A check is conducted to verify that the culvert velocity is less than or equal to at least part of the upstream or downstream channel. Table J.12 summarizes the velocities estimated at each cross-section and within the culvert by the HY-8/Normal Depth and HEC-RAS methods. The check is satisfied if the culvert inlet and outlet velocities are within the range of the cross-section velocities. For the remaining computations, the check is satisfied. Proceed to Step 11.

Table J.12. Velocity Estimates at Q_H.

Cross-section**	Applicable Reach Length (ft)	Normal Depth/HY-8 (ft/s)	HEC-RAS (ft/s)
875	52	4.17	2.93
823	51	4.41	3.96
772	53	6.43	4.47
Culvert Inlet*	18	4.78	3.71
Culvert Outlet*	18	4.93	2.56
666	52	3.93	2.71
614	64	4.25	3.78
550	80	4.02	4.46
470	50	4.54	4.49

*10.0-ft CMP, embedment=3.0 ft, S_o=0.007 ft/ft, n_{bed}=0.016.
**n_{bed} = 0.016

Step 11. Check Culvert Water Depth at Q_L.

A check is conducted to verify that the culvert depth is greater than or equal to at least part of the upstream or downstream channel. Field data for the low flow channels were not available for cross-sections 666 through 470. For this reason, the depths in these cross-sections may be underestimated. Table J.13 summarizes the depth calculations.

Comparing the upstream and downstream water depths to the water depths within the culvert, we see the depth at the culvert inlet is the shallowest in the project reach. Therefore, we will proceed to Step 12 to create a low flow channel.

Table J.13. Maximum Depth Estimates at Q_L.

Cross-section**	Normal Depth/HY-8 (ft)	HEC-RAS (ft)
875	0.40	0.40
823	0.30	0.47
772	0.23	0.35
Culvert Inlet*	0.10	0.14
Culvert Outlet*	0.19	0.25
666	0.22	0.27
614	0.18	0.35
550	0.15	0.45
470	0.18	0.39

*10.0-ft CMP, embedment=3.0 ft, S_o=0.007 ft/ft, n_{bed}=0.028.
**n_{bed} = 0.028

Step 12. Provide Low-flow Channel in Culvert.

To increase the depth in the culvert bed add a triangular low-flow channel with side slopes of 1:8 (V:H). This will provide a thalweg 0.6 ft deeper in the center of the culvert within the upper bed layer of sand. It is expected that the stream will likely rearrange the low flow channel. It may also be possible to forego creation of the low flow channel at construction and allowing the stream to carve the low flow channel.

Step 13. Review Design.

A 10.0-ft (3.05-m) CMP with a 3-ft (0.91-m) embedment on a 0.7 percent slope is proposed to replace the twin 3.0-ft (0.91-m) CMP culverts on a 1.5 percent slope. An oversized bed is to be placed in the culvert below a layer of native bed material to provide stability at Q_P. A low-flow channel is to be created to maintain depths in the culvert at Q_L.

Alternative culvert shapes and materials may also be considered. A concrete box or pipe arch may offer an option to maintain a sufficiently wide span to meet the stability, velocity, and depth criteria with a lower rise.

J.3 SUPPORTING DOCUMENTATION

J.3.1 Cross-sections.

J.3.2 HY-8 Report for 7.0-ft CMP at Q_H.

J.3.3 HEC-RAS Output for 7.0-ft CMP at Q_H.

J.3.4 HY-8 Report for 7.0-ft CMP with Oversize Bed Material at Q_P.

J.3.5 HEC-RAS Output for 7.0-ft CMP with Oversize Bed Material at Q_P.

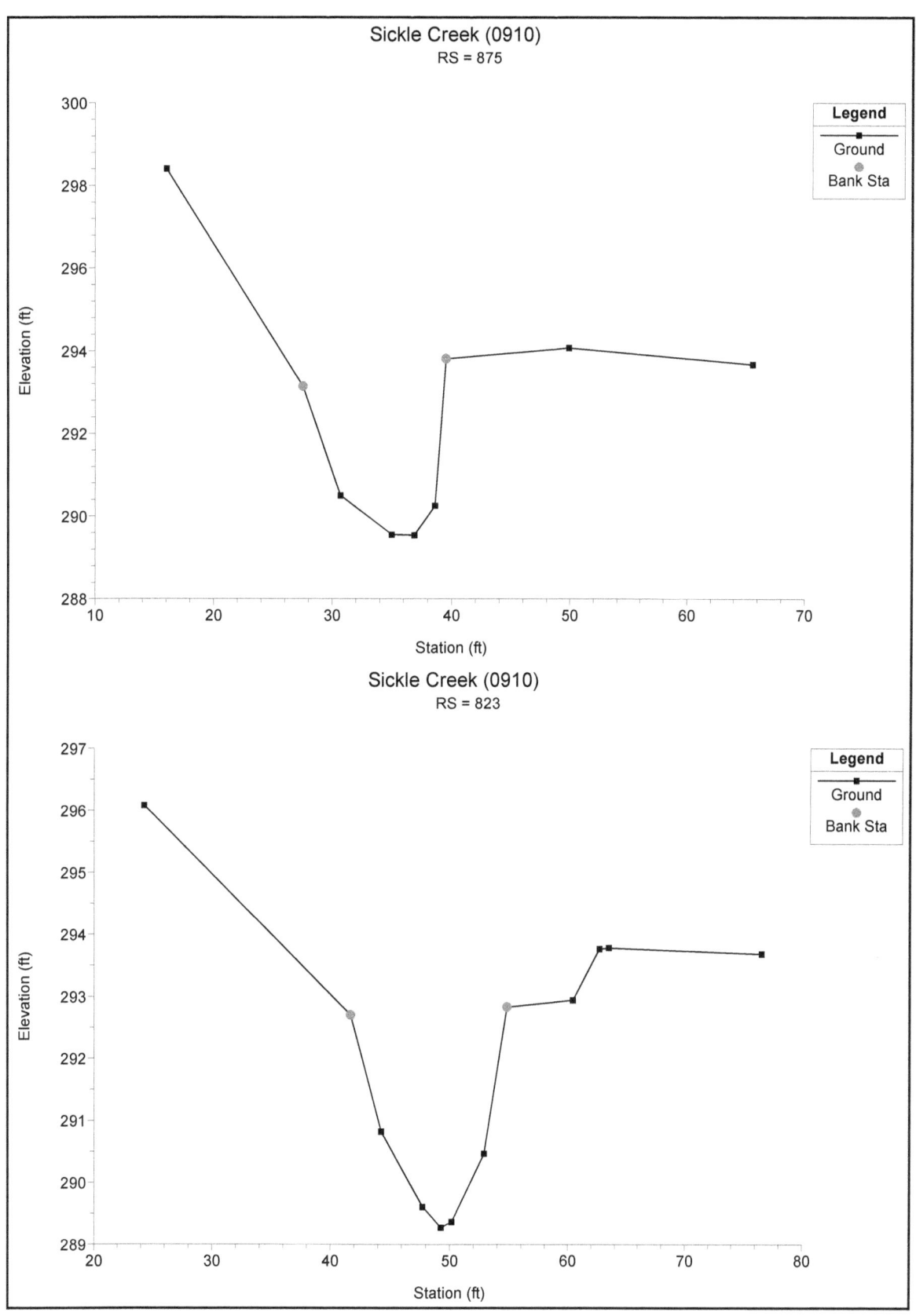

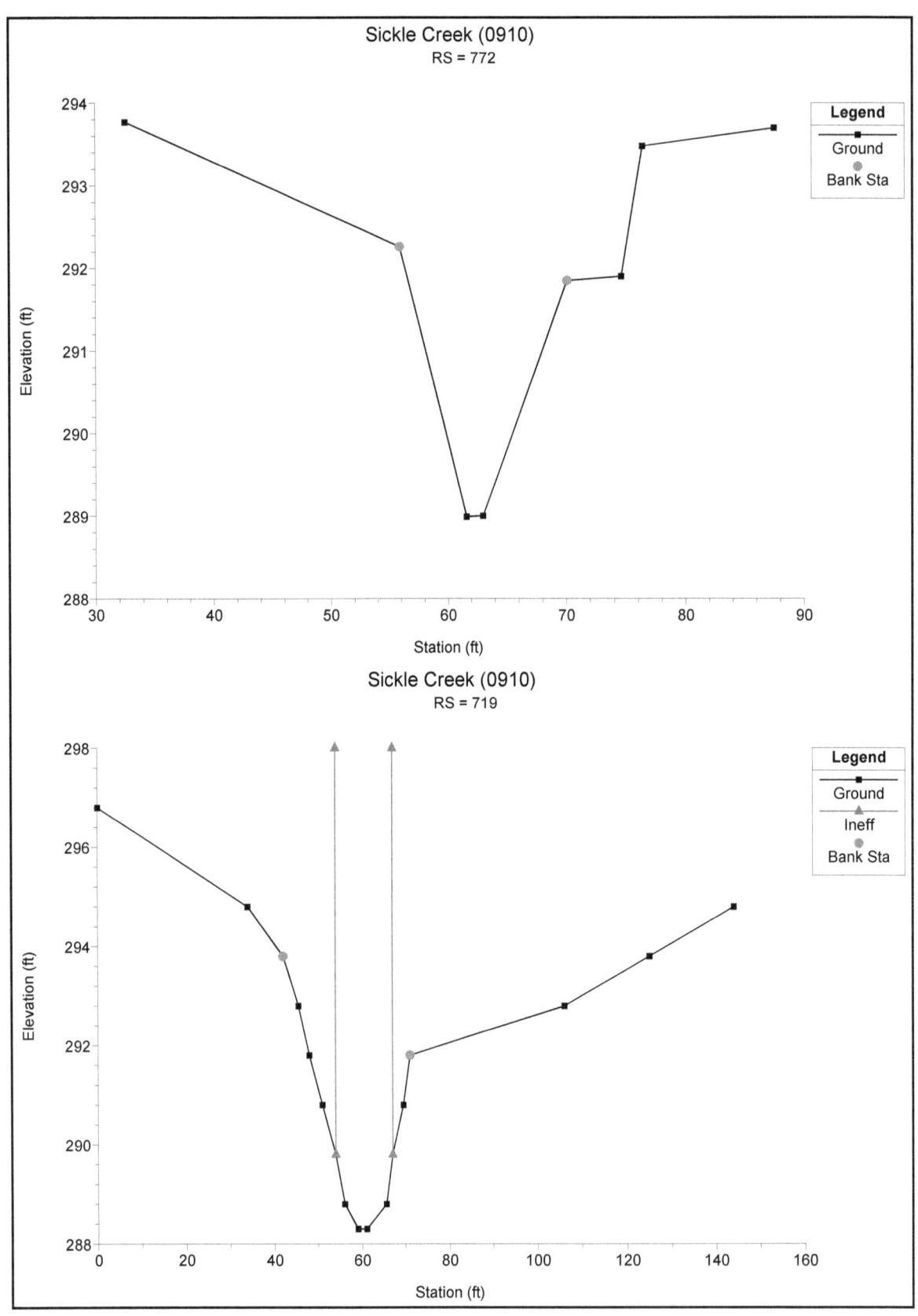

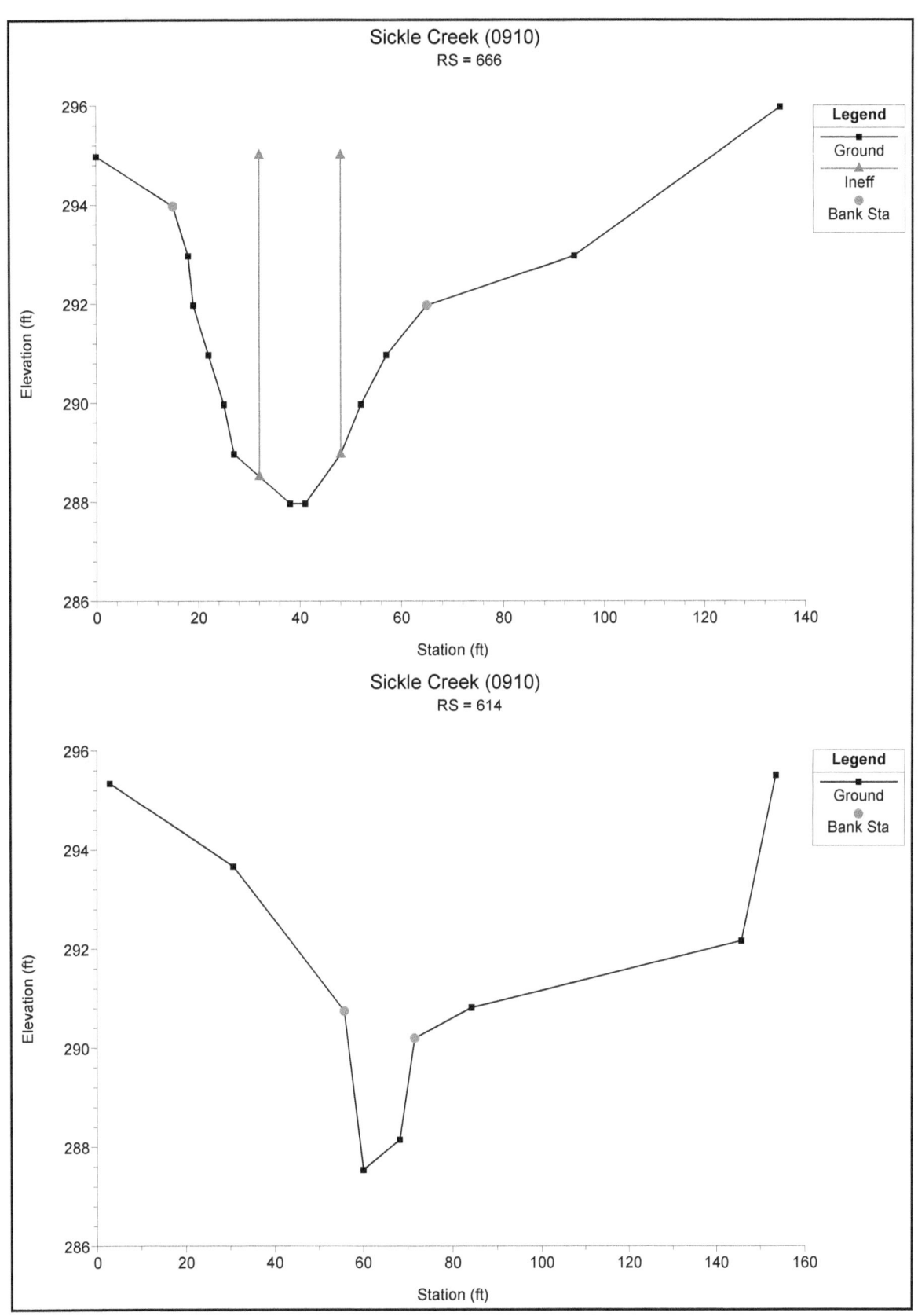

Sickle Creek (0910)
RS = 666

Sickle Creek (0910)
RS = 614

J-15

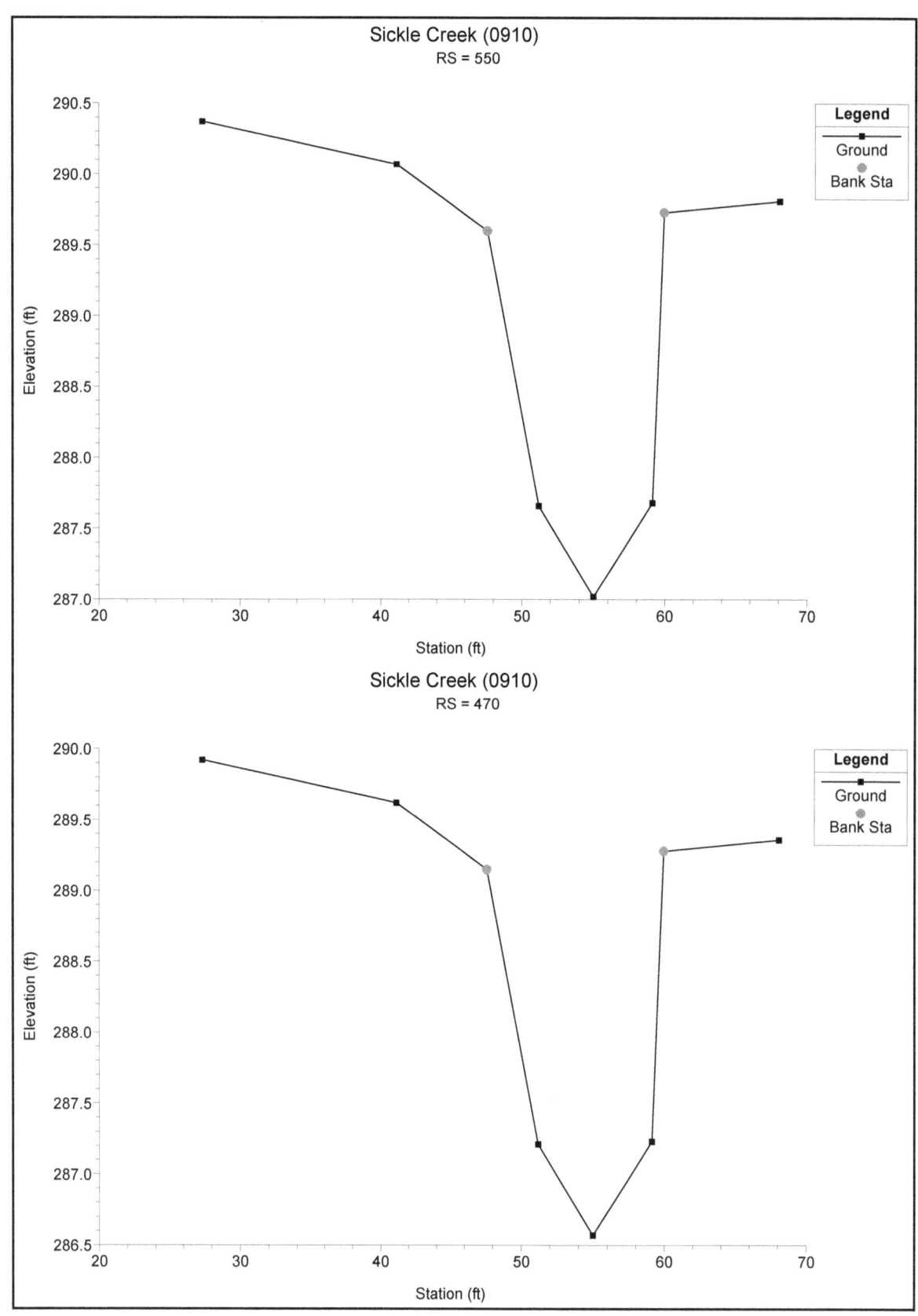

Table 1 - Culvert Summary Table: 7.0' CMP emb=2.1' n= 0.016

Total Discharge (cfs)	Culvert Discharge (cfs)	Headwater Elevation (ft)	Inlet Control Depth (ft)	Outlet Control Depth (ft)	Flow Type	Normal Depth (ft)	Critical Depth (ft)	Outlet Depth (ft)	Tailwater Depth (ft)	Outlet Velocity (ft/s)	Tailwater Velocity (ft/s)
0.00	0.00	288.25	0.000	0.0*	0-NF	0.000	0.000	0.000	0.000	0.000	0.000
15.00	15.00	289.11	0.859	0.162	1-S2n	0.503	0.542	0.532	0.442	4.733	4.869
32.60	32.60	289.67	1.422	0.351	1-S2n	0.799	0.907	0.806	0.631	6.087	5.960
48.90	48.90	290.10	1.852	0.475	1-S2n	1.054	1.179	1.069	0.755	6.818	6.614
65.20	65.20	290.48	2.233	0.576	1-S2n	1.270	1.434	1.287	0.856	7.505	7.118
81.50	81.50	290.83	2.584	0.662	1-S2n	1.487	1.650	1.530	0.942	7.851	7.534
97.80	97.80	291.17	2.924	0.736	1-S2n	1.708	1.860	1.750	1.016	8.212	7.925
114.10	114.10	291.51	3.265	0.797	1-S2n	1.928	2.054	1.933	1.077	8.657	8.369
130.40	130.40	291.84	3.594	0.854	1-S2n	2.151	2.234	2.224	1.134	8.603	8.769
146.70	146.70	292.13	3.884	0.908	1-S2n	2.374	2.414	2.374	1.188	9.070	9.135
163.00	163.00	292.80	4.174	4.554	2-M2c	2.608	2.575	2.582	1.240	9.278	9.473

****

Inlet Elevation (invert): 288.25 ft,   Outlet Elevation (invert): 288.00 ft

Culvert Length: 36.00 ft,   Culvert Slope: 0.0069

****

Site Data - 7.0' CMP emb=2.1' n= 0.016

Site Data Option: Culvert Invert Data

Inlet Station: 0.00 ft

Inlet Elevation: 286.15 ft

Outlet Station: 36.00 ft

Outlet Elevation: 285.90 ft

Number of Barrels: 1

Culvert Data Summary - 7.0' CMP emb=2.1' n= 0.016

Barrel Shape: Circular

Barrel Diameter: 7.00 ft

Barrel Material: Corrugated Steel

Embedment: 25.20 in

Barrel Manning's n: 0.0240 (top and sides)

Manning's n: 0.0160 (bottom)

Inlet Type: Conventional

Inlet Edge Condition: Square Edge with Headwall

Inlet Depression: None

Table 2 - Downstream Channel Rating Curve (Crossing: Sickle Creek (A), n=0.016)

Flow (cfs)	Water Surface Elev (ft)	Depth (ft)	Velocity (ft/s)	Shear (psf)	Froude Number
0.00	287.97	0.00	0.00	0.00	0.00
15.00	288.41	0.44	4.87	0.41	1.62
32.60	288.60	0.63	5.96	0.59	1.70
48.90	288.72	0.75	6.61	0.71	1.75
65.20	288.83	0.86	7.12	0.80	1.78
81.50	288.91	0.94	7.53	0.88	1.80
97.80	288.99	1.02	7.93	0.95	1.83
114.10	289.05	1.08	8.37	1.01	1.85
130.40	289.10	1.13	8.77	1.06	1.87
146.70	289.16	1.19	9.14	1.11	1.89
163.00	289.21	1.24	9.47	1.16	1.91

Tailwater Channel Data - Sickle Creek (A), n=0.016

    Tailwater Channel Option: Irregular Channel

Roadway Data for Crossing: Sickle Creek (A), n=0.016

    Roadway Profile Shape: Irregular Roadway Shape (coordinates)

    Roadway Surface: Gravel

    Roadway Top Width: 24.00 ft

J.3.3. HEC-RAS Output for 7.0-ft CMP at QH.

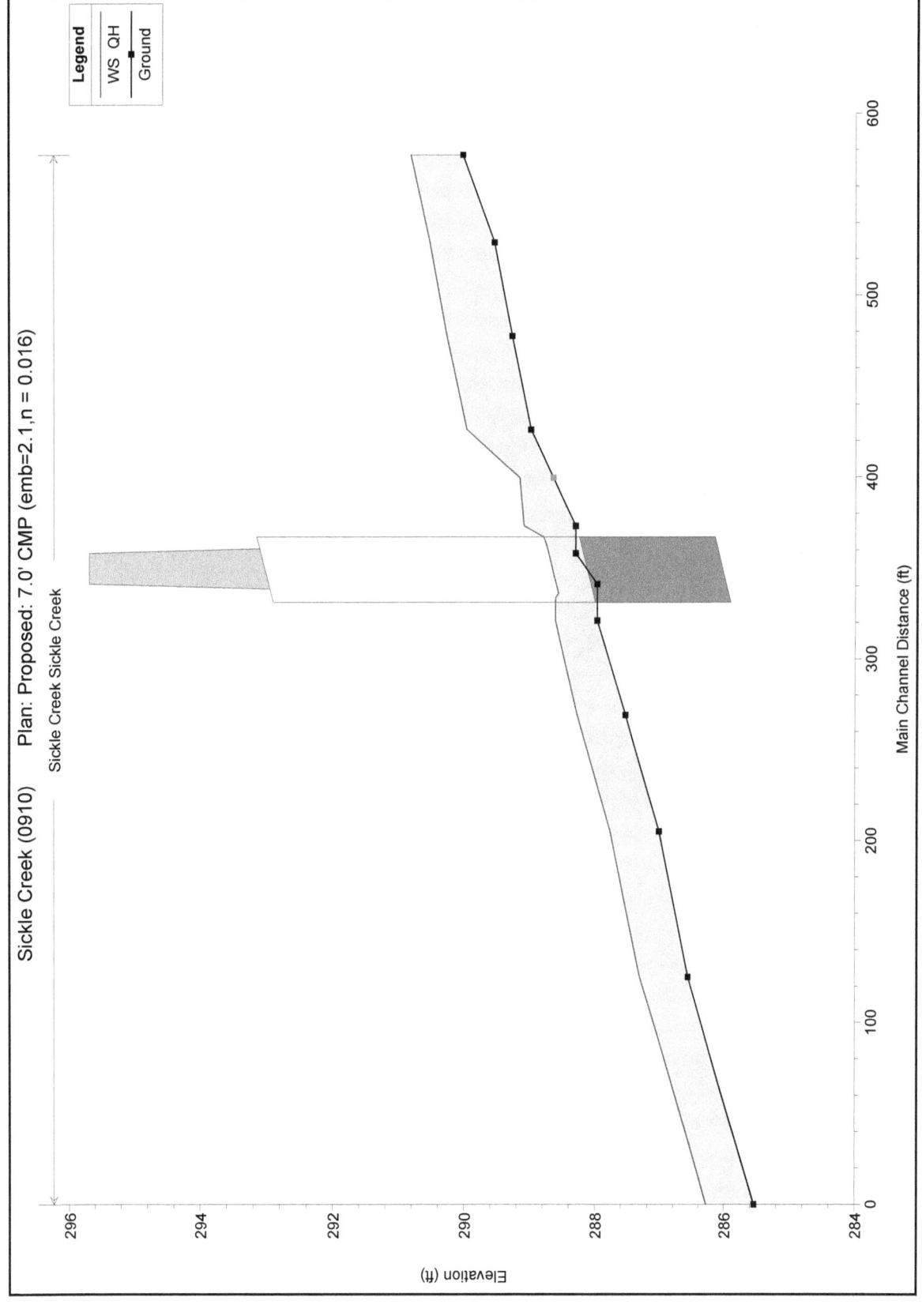

J.3.3. HEC-RAS Output for 7.0-ft CMP at QH.

HEC-RAS Plan: 7.0' 2.1' 0.016   River: Sickle Creek   Reach: Sickle Creek   Profile: QH

Reach	River Sta	Profile	Q Total (cfs)	Min Ch El (ft)	W.S. Elev (ft)	Max Chl Dpth (ft)	Hydr Depth (ft)	Mann Wtd Chnl	Crit W.S. (ft)	E.G. Elev (ft)	E.G. Slope (ft/ft)	Vel Chnl (ft/s)	Flow Area (sq ft)	Top Width (ft)	Froude # Chl
Sickle Creek	923	QH	15.00	290.02	290.82	0.80	0.51	0.016	290.82	291.08	0.005036	4.09	3.66	7.19	1.01
Sickle Creek	875	QH	15.00	289.54	290.53	0.99	0.64	0.016	290.34	290.66	0.001943	2.93	5.13	8.01	0.64
Sickle Creek	823	QH	15.00	289.27	290.27	1.00	0.58	0.016	290.23	290.51	0.004050	3.96	3.79	6.57	0.92
Sickle Creek	772	QH	15.00	288.99	289.97	0.98	0.61	0.016	289.97	290.28	0.004982	4.47	3.36	5.51	1.01
Sickle Creek	745.5*	QH	15.00	288.64	289.16	0.52	0.34	0.016	289.40	289.96	0.026114	7.20	2.08	6.14	2.18
Sickle Creek	719	QH	15.00	288.30	289.09	0.79	0.55	0.016	288.93	289.20	0.001761	2.58	5.81	10.52	0.61
Sickle Creek	695	Culvert													
Sickle Creek	666	QH	15.00	287.97	288.61	0.64	0.41	0.016		288.72	0.002811	2.71	5.53	14.49	0.75
Sickle Creek	614	QH	15.00	287.53	288.28	0.75	0.42	0.016	288.28	288.49	0.005355	3.76	3.99	9.41	1.02
Sickle Creek	550	QH	15.00	287.02	287.77	0.75	0.41	0.016	287.82	288.08	0.007895	4.48	3.35	8.17	1.23
Sickle Creek	470	QH	15.00	286.57	287.31	0.74	0.41	0.016	287.37	287.63	0.008182	4.53	3.31	8.16	1.25
Sickle Creek	344	QH	15.00	285.54	286.28	0.74	0.40	0.016	286.34	286.60	0.008354	4.56	3.29	8.15	1.27

J.3.3. HEC-RAS Output for 7.0-ft CMP at QH.

Plan: 7.0' 2.1' 0.016   Sickle Creek   Sickle Creek RS: 695   Culv Group: Culvert 1   Profile: QH

Q Culv Group (cfs)	15.00	Culv Full Len (ft)	
# Barrels	1	Culv Vel US (ft/s)	4.14
Q Barrel (cfs)	15.00	Culv Vel DS (ft/s)	3.72
E.G. US. (ft)	289.20	Culv Inv El Up (ft)	286.15
W.S. US. (ft)	289.09	Culv Inv El Dn (ft)	285.90
E.G. DS (ft)	288.72	Culv Frctn Ls (ft)	0.24
W.S. DS (ft)	288.61	Culv Exit Loss (ft)	0.10
Delta EG (ft)	0.47	Culv Entr Loss (ft)	0.13
Delta WS (ft)	0.48	Q Weir (cfs)	
E.G. IC (ft)	289.05	Weir Sta Lft (ft)	
E.G. OC (ft)	289.20	Weir Sta Rgt (ft)	
Culvert Control	Outlet	Weir Submerg	
Culv WS Inlet (ft)	288.80	Weir Max Depth (ft)	
Culv WS Outlet (ft)	288.61	Weir Avg Depth (ft)	
Culv Nml Depth (ft)	2.63	Weir Flow Area (sq ft)	
Culv Crt Depth (ft)	2.65	Min El Weir Flow (ft)	296.45

Errors Warnings and Notes

Note:	During supercritical analysis, the culvert direct step method went to normal depth. The program then assumed normal depth at the outlet.
Note:	During the supercritical calculations a hydraulic jump occurred inside of the culvert.

Table 1 - Culvert Summary Table: 10.0' CMP emb=2.0' n= 0.052

Total Discharge (cfs)	Culvert Discharge (cfs)	Headwater Elevation (ft)	Inlet Control Depth (ft)	Outlet Control Depth (ft)	Flow Type	Normal Depth (ft)	Critical Depth (ft)	Outlet Depth (ft)	Tailwater Depth (ft)	Outlet Velocity (ft/s)	Tailwater Velocity (ft/s)
0.00	0.00	287.97	0.000	0.720	0-NF	0.000	0.000	0.000	0.000	0.000	0.000
16.30	16.30	288.61	0.572	1.355	3-M1t	0.915	0.394	1.495	0.525	1.239	4.017
32.60	32.60	288.97	1.163	1.717	3-M1t	1.410	0.789	1.687	0.717	2.175	4.805
48.90	48.90	289.29	1.522	2.042	3-M2t	1.829	0.996	1.827	0.857	2.994	5.330
65.20	65.20	289.60	1.826	2.345	3-M2t	2.200	1.198	1.940	0.970	3.741	5.734
81.50	81.50	289.87	2.105	2.624	3-M2t	2.543	1.400	2.028	1.058	4.455	6.161
97.80	97.80	290.14	2.366	2.895	3-M2t	2.852	1.601	2.105	1.135	5.134	6.567
114.10	114.10	290.41	2.606	3.158	3-M2t	3.162	1.749	2.177	1.207	5.776	6.928
130.40	130.40	290.66	2.835	3.411	3-M2t	3.421	1.898	2.244	1.274	6.386	7.253
146.70	146.70	290.90	3.054	3.655	3-M2t	3.673	2.046	2.308	1.338	6.970	7.550
163.00	163.00	291.14	3.264	3.892	3-M2t	3.926	2.194	2.368	1.398	7.531	7.823

****

Inlet Elevation (invert): 287.25 ft,   Outlet Elevation (invert): 287.00 ft

Culvert Length: 36.00 ft,   Culvert Slope: 0.0069

****

Site Data - 10.0' CMP emb=2.0' n= 0.052

Site Data Option: Culvert Invert Data

Inlet Station: 0.00 ft

Inlet Elevation: 285.25 ft

Outlet Station: 36.00 ft

Outlet Elevation: 285.00 ft

Number of Barrels: 1

Culvert Data Summary - 10.0' CMP emb=2.0' n= 0.052

Barrel Shape: Circular

Barrel Diameter: 10.00 ft

Barrel Material: Corrugated Steel

Embedment: 24.00 in

Barrel Manning's n: 0.0240 (top and sides)

Manning's n: 0.0520 (bottom)

Inlet Type: Conventional

Inlet Edge Condition: Square Edge with Headwall

Inlet Depression: None

Table 2 - Downstream Channel Rating Curve (Crossing: Sickle Creek (C-QP), n=0.016)

Flow (cfs)	Water Surface Elev (ft)	Depth (ft)	Velocity (ft/s)	Shear (psf)	Froude Number
0.00	287.97	0.00	0.00	0.00	0.00
16.30	288.50	0.53	4.02	0.28	1.24
32.60	288.69	0.72	4.80	0.38	1.30
48.90	288.83	0.86	5.33	0.45	1.33
65.20	288.94	0.97	5.73	0.51	1.36
81.50	289.03	1.06	6.16	0.55	1.38
97.80	289.11	1.14	6.57	0.59	1.40
114.10	289.18	1.21	6.93	0.63	1.42
130.40	289.24	1.27	7.25	0.67	1.43
146.70	289.31	1.34	7.55	0.70	1.45
163.00	289.37	1.40	7.82	0.73	1.46

Tailwater Channel Data - Sickle Creek (C-QP), n=0.016

Tailwater Channel Option: Irregular Channel

Roadway Data for Crossing: Sickle Creek (C-QP), n=0.016

Roadway Profile Shape: Irregular Roadway Shape (coordinates)

Roadway Surface: Gravel

Roadway Top Width: 24.00 ft

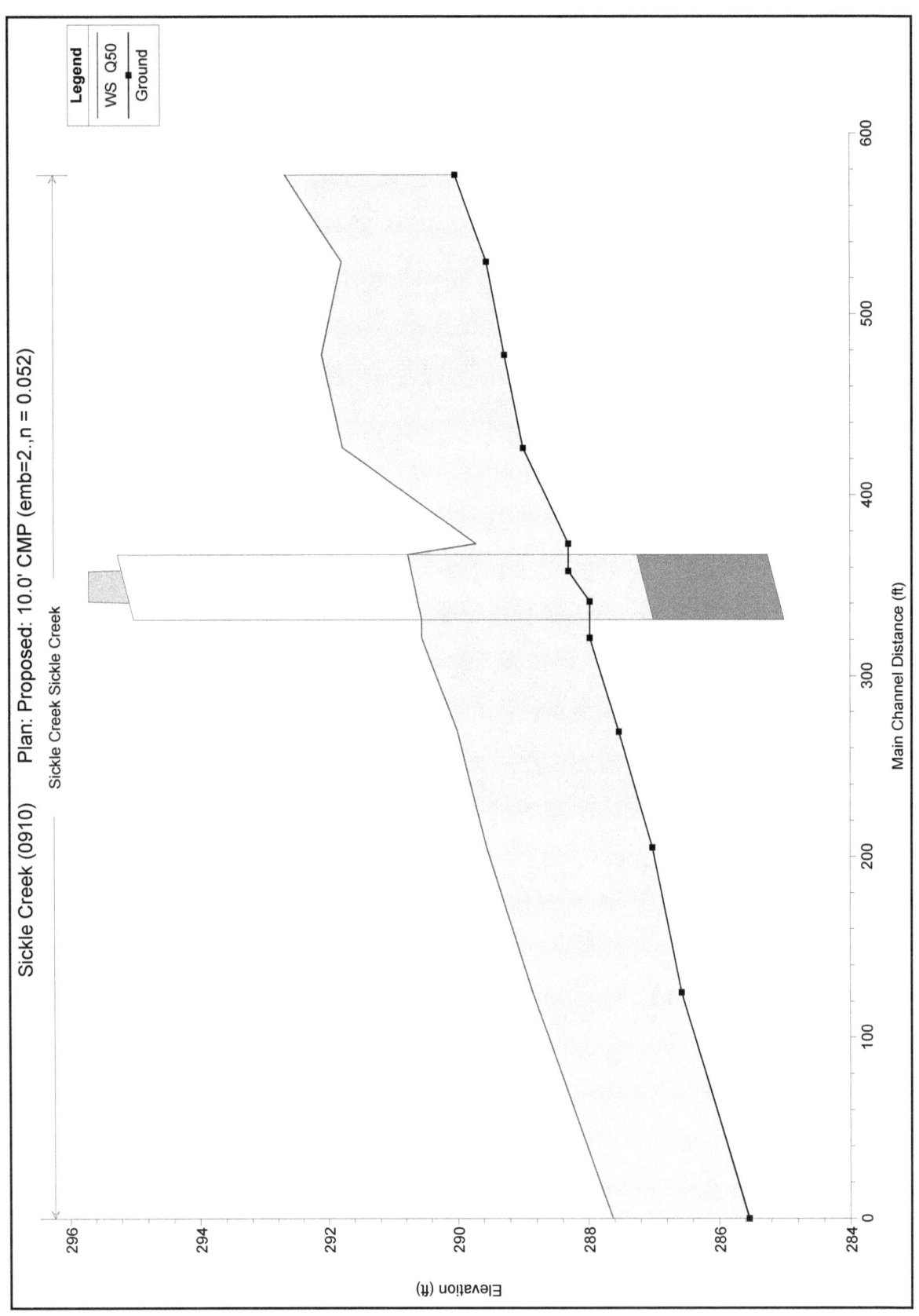

Sickle Creek (0910)   Plan: Proposed: 10.0' CMP (emb=2.,n = 0.052)

Sickle Creek Sickle Creek

Legend

WS Q50

Ground

Main Channel Distance (ft)

Elevation (ft)

J-28

HEC-RAS Plan: 10' 2' 0.052   River: Sickle Creek   Reach: Sickle Creek   Profile: Q50

Reach	River Sta	Profile	Q Total (cfs)	Min Ch El (ft)	W.S. Elev (ft)	Max Chl Dpth (ft)	Hydr Depth (ft)	Mann Wtd Chnl	Crit W.S. (ft)	E.G. Elev (ft)	E.G. Slope (ft/ft)	Vel Chnl (ft/s)	Flow Area (sq ft)	Top Width (ft)	Froude # Chl
Sickle Creek	923	Q50	163.00	290.02	292.66	2.64	1.95	0.016	292.67	293.66	0.003981	8.00	20.39	10.43	1.01
Sickle Creek	875	Q50	163.00	289.54	291.78	2.24	1.65	0.016	292.19	293.34	0.007559	10.03	16.24	9.83	1.38
Sickle Creek	823	Q50	163.00	289.27	292.08	2.81	1.81	0.016	292.09	293.01	0.003729	7.72	21.12	11.66	1.01
Sickle Creek	772	Q50	163.00	288.99	291.77	2.78	1.53	0.016	292.00	292.78	0.004876	8.10	20.11	13.13	1.15
Sickle Creek	719	Q50	163.00	288.30	289.71	1.41	1.02	0.016	290.40	292.18	0.019103	12.63	12.91	12.67	2.20
Sickle Creek	695		Culvert												
Sickle Creek	666	Q50	163.00	287.97	290.54	2.57	2.25	0.016		290.86	0.000813	4.53	36.00	31.57	0.53
Sickle Creek	614	Q50	163.00	287.53	289.99	2.46	1.69	0.016	289.86	290.68	0.002851	6.63	24.58	14.53	0.90
Sickle Creek	550	Q50	163.00	287.02	289.55	2.53	1.76	0.016	289.55	290.44	0.003784	7.59	21.48	12.19	1.01
Sickle Creek	470	Q50	163.00	286.57	288.86	2.29	1.60	0.016	289.10	290.05	0.005651	8.75	18.63	11.65	1.22
Sickle Creek	344	Q50	163.00	285.54	287.63	2.09	1.46	0.016	288.07	289.17	0.008159	9.96	16.36	11.20	1.45

Plan: 10' 2' 0.052   Sickle Creek   Sickle Creek RS: 695   Culv Group: Culvert 1   Profile: Q50

Q Culv Group (cfs)	163.00	Culv Full Len (ft)	
# Barrels	1	Culv Vel US (ft/s)	4.92
Q Barrel (cfs)	163.00	Culv Vel DS (ft/s)	4.87
E.G. US. (ft)	292.18	Culv Inv El Up (ft)	285.25
W.S. US. (ft)	289.71	Culv Inv El Dn (ft)	285.00
E.G. DS (ft)	290.86	Culv Frctn Ls (ft)	0.22
W.S. DS (ft)	290.54	Culv Exit Loss (ft)	0.05
Delta EG (ft)	1.32	Culv Entr Loss (ft)	0.19
Delta WS (ft)	0.84	Q Weir (cfs)	
E.G. IC (ft)	290.49	Weir Sta Lft (ft)	
E.G. OC (ft)	291.32	Weir Sta Rgt (ft)	
Culvert Control	Outlet	Weir Submerg	
Culv WS Inlet (ft)	290.76	Weir Max Depth (ft)	
Culv WS Outlet (ft)	290.54	Weir Avg Depth (ft)	
Culv Nml Depth (ft)	5.40	Weir Flow Area (sq ft)	
Culv Crt Depth (ft)	4.20	Min El Weir Flow (ft)	296.45

Errors Warnings and Notes

Note:	During the supercritical calculations a hydraulic jump occurred at the inlet of (going into) the culvert.

J.4 REFERENCES

DeBoer, Jason, Kris N. Nault, Eric B. Snyder, J. Marty Holtgren and Stephanie Ogren, 2007. "Fish Response to Habitat Restoration on Sickle Creek, a First Order Tributary of the Big Manistee River," presented at the North American Benthological Society 55th Annual Meeting, June 5, 2007.

Kilgore, Roger T. and George K. Cotton, 2005. "Design of Roadside Channels with Flexible Linings," Hydraulic Engineering Circular Number 15 (HEC 15), 3rd Edition, FHWA-NHI-05-114, Federal Highway Administration.

Michigan Department of Transportation, 2006. "Drainage Manual," January.

Holtschlag, D.J. and Hope M. Croskey, 1984. "Statistical Models for Estimating Flow Characteristics of Michigan Streams," U.S.G.S. Water-resources Investigations Report 84-4207.

Ogren, Stephanie, Marty Holtgren, Pat Fowler, and Mike Joyce, 2008. "Sickle Creek Case Study," http://stream.fs.fed.us/fishxing/case/sickle_creek/index.html accessed November 18.

Sorrell, Richard C., 2008. "Computing Flood Discharges for Small Ungaged Watersheds," Michigan Department of Environmental Quality, Land and Water Management Division, June.

This page intentionally left blank.

APPENDIX K- DESIGN EXAMPLE RESULTS COMPARISON

The design examples presented in Appendices H, I, and J are based on actual AOP culvert replacement projects. In each case, a replacement structure was constructed based on an alternative geomorphic design procedure. The following table compares the design result using this HEC 26 procedure with the passage structure as built. The table also provides the size and type of structure that was previously located at the site creating the AOP barrier.

It should be noted that the design examples all produced a design using a round embedded culvert. The designer could have evaluated other embedded shapes (arches, squash pipes, and boxes) or open-bottomed culverts.

Table K.1. Structure Comparisons for Three Case Studies.

	North Thompson (Appendix H)	Trib to Bear Creek (Appendix I)	Sickle Creek (Appendix J)
AOP barrier	3-ft CMP	5-ft CMP	Twin 3-ft CMPs
As-built	12'x ? squash pipe	9.75'x 6.6' pipe arch	16'x 6' concrete arch bridge
HEC-26 procedure	8.5' CMP	12' CMP	10' CMP
Difference in span	-3.5 ft	+2.25 ft	-6 ft

This page intentionally left blank.

www.ingramcontent.com/pod-product-compliance
Lightning Source LLC
Chambersburg PA
CBHW080636180526
45168CB00008B/3193